The Invention of Order

RADICAL AMÉRICAS
A series edited by Bruno Bosteels and Geo Maher

The Invention of Order

On the Coloniality of Space | DON THOMAS DEERE

Foreword by Santiago Castro-Gómez

Duke University Press *Durham and London* 2025

Typeset in Garamond Premier Pro by Westchester Publishing
Services

Library of Congress Cataloging-in-Publication Data
Names: Deere, Don Thomas, [date] author.
Title: The invention of order : on the coloniality of space /
Don Thomas Deere.
Other titles: Radical Américas.
Description: Durham : Duke University Press, 2025. |
Series: Radical Américas | Includes bibliographical
references and index.
Identifiers: LCCN 2025020362 (print)
LCCN 2025020363 (ebook)
ISBN 9781478032878 (paperback)
ISBN 9781478029427 (hardcover)
ISBN 9781478061625 (ebook)
Subjects: LCSH: Cartography—Political aspects. | Settler
colonialism—America. | Imperialism. | Postcolonialism. |
Human geography—Political aspects. | America—Discovery
and exploration—Maps.
Classification: LCC G1101.S12 D44 2025 (print) | LCC G1101.
S12 (ebook) | DDC 306.2097—dc23/eng/20250616
LC record available at https://lccn.loc.gov/2025020362
LC ebook record available at https://lccn.loc.gov/2025020363

Cover art: *Monumento*, c. 1944. 53 × 39¼ × 2 in. Courtesy of
the Estate of Joaquín Torres-García. Photo by Arturo Sanchez.

Contents

Foreword

SANTIAGO CASTRO-GÓMEZ

During the last decades of the twentieth century, one began to speak of the spatial turn. The displacement of the time-centered approach that had dominated the modern social sciences since the eighteenth century turned toward another focus centered on space. The concept of space is viewed, here, not merely as a natural container for the passage of history but rather as an actor capable of redefining the social relations of power. Geography, which had emerged as a science in the nineteenth century, was driven by the development of other areas of study such as urban sociology to begin to understand space as a fundamental variable for comprehending historical processes and inequality between diverse territories.

In the Anglo-American context, particular attention was accorded to the work of the Marxist geographer David Harvey, for whom struggles for control of land and territory are fundamental to the functioning of global capitalism. The tendency toward the incessant accumulation of capital drives a constant reorganization of space, which can be observed in the continuous expansion of cities, the construction of infrastructures, and the creation of new forms of deterritorialized work. His book *The Condition of Postmodernity* shows how capitalism contributes to the creation of a new geography in urban and global peripheries, such as the increase in suburbs and slums occupied by those who are homeless, unemployed, or ethnic minorities. What is interesting about the work of Harvey is that he presents geography not as a descriptive or neutral science but as a weapon to transform existing social relations in an emancipatory sense. Geography, in other words, functions as a counterscience.

In the field of philosophy, one must certainly mention the work of Michel Foucault, according to whom spaces exist as laboratories for generating social behaviors: spaces such as hospitals, prisons, schools, and factories, which not only physically organize people but also mold their behaviors. With the image of the panopticon, Foucault shows how spaces can internalize and naturalize relations of power. While examining space as a tool of control, however, Foucault was also interested in space as a tool of resistance. He speaks of "heterotopias," which are spaces of alterity and resistance in the face of power, where it is possible to defy the established norms. Critical theories of space in the first world have developed much from this idea, showing how social movements and actions of civil disobedience use space as sites by which to defy and subvert the structures of established power. Space is converted into a site of symbolic and material struggle against structures of power, as in the protests of the Occupy movement. Such is the case of the Indignados in Spain, the yellow vests in France, and Occupy Wall Street in the United States, for whom public space is resignified and utilized for resistance.

Yet all these theories of space leave out a key element for the comprehension of the reconfiguration of space proper to modernity. I am referring to the role of the colonization of the Americas in the sixteenth century, which operated as a prior condition as much for the configuration of capitalism of which Harvey speaks as it does for the disciplinary society of which Foucault speaks. This point has been highlighted since the beginning of the twenty-first century by the decolonial theories of Latin America. Authors such as Aníbal Quijano, Enrique Dussel, María Lugones, and Arturo Escobar have emphasized how racial, ethnic, and gender identities are interwoven with space. Indigenous communities, people of African descent, and mestizas have had to inhabit spaces that not only segregated them physically but also imposed colonial identities on them. The reconfiguration of these spaces, thus, is also a struggle for the identity and cultural autonomy of such populations.

It is precisely here that this book of Don Deere, in your hands, takes on its significance. The concept that gives his book its title, the coloniality of space, not only corrects the Eurocentrism of Harvey and Foucault but amplifies and deepens the notion of the coloniality of power coined by Quijano. The colonial matrix of power produced in the Americas during the sixteenth century includes not only the racial classification of populations vis-à-vis its role in the international division of labor—as Quijano says—but also the organization and hierarchization of physical space. Deere, in other words, goes beyond Quijano in asking after the spatial conditions of social classification. His thesis is that it was possible to control and organize colonial settlements only through the creation of spatial grids, shaping also the hierarchization and classification of racialized populations. These grids operate as a kind of spatial design reflecting the power and administrative necessity of the colonizers. They were constructed by refortifying social hierarchies and racial segregation, removing Indigenous populations from their own territory and imposing on them new forms of relating to natural resources and to their own traditions.

One of the central objectives of the book is to present a spatial reading of modernity that takes into account its structural connections with coloniality. This goal positions Deere as not only an inheritor of decolonial theory but also a philosopher capable of proposing new interpretations of now canonical authors in the discipline, such as Foucault. His thesis that the American colonies of the sixteenth century functioned as laboratories to experiment with the two modern technologies of power that Foucault referred to as anatomo-politics and biopolitics in *The History of Sexuality, Volume 1* is striking. As is well known, Foucault dates the birth of these technologies to the eighteenth century: first anatomo-politics, oriented toward the discipline of bodies, and then biopolitics, oriented toward the governance of populations. Yet all of this, according to Foucault, occurred in Europe. Instead, Deere argues that both technologies of power are products of the asymmetrical interaction between colonizers and the colonized, which occurred outside Europe, in the American colonies of the sixteenth century. It was only after, in the manner of a boomerang effect, that those technologies were applied in the eighteenth century to European bodies and populations. The point is that these technologies (the hospital, the factory, etc.) would not have emerged in the intra-European spaces marked by Foucault if they had not been used beforehand to control and govern Indigenous and Black populations.

Colonial spaces imposed a form of relating to territory that contrasted with the developed urban models in pre-Columbian societies. Coloniality redefined space in terms of its signification and utilization, assuring European domination over conquered territories and imposing over them a new narrative. The imposition of urban grids functioned as mechanisms of surveillance and control, long before the disciplinary societies of which Foucault speaks emerged in Europe. Spatial practices of the colonizers reinforced the narrative that presented the Europeans as civilized and the colonized as barbarians. The new spatial organization justified narratives of European superiority while simultaneously delegitimizing forms of knowledge and life proper to local communities. In this sense, Deere examines how the ordering of space tested in the colonies was then converted into the model of domination implemented within Europe itself. With this move, Deere destabilizes the Eurocentric theories of space advanced by Foucault and Harvey and at the same time complements and strengthens Quijano's theory.

One of the key moments of the book is the discussion that Deere establishes with Carl Schmitt's *The Nomos of the Earth*. The German philosopher is correct to argue that modernity begins with the surge of the maritime empires of the Atlantic (Spain and Portugal) and that this phenomenon presupposes the creation of a new nomos that globally codifies the sea as much as the earth. This is the modern *ius gentium*, which finds its first great theorist in Francisco de Vitoria. Schmitt does not tire of praising Vitoria for having recognized that the old order of space, the medieval *res publica Christiana*, has ceased to be operative and that now a new order that goes beyond European Christianity is required; a new order in which Christians as much as non-Christians possess equal right to traverse the globe. Deere points out, however, that while this new nomos gives to Europeans the right to liberal commerce and to limitlessly propagate their religious ideas, it takes away the right of autodetermination of Indigenous peoples vis-à-vis their own cultures and territories. If Indigenous peoples opposed free commerce and the purported Christianization offered by the Europeans, they thereby violated international law, thus providing a motive to wage a just war against them and appropriate their lands. The new nomos, Deere affirms, operates thus as an emptying of space and functions as the perfect legitimization of the primitive accumulation of capital identified by Karl Marx.

However, the history of modernity cannot be seen solely as the history of colonial power. Foucault himself already showed that power must be understood as a conjunction of relations that can never be closed, that always leave gaps, cracks, and spaces of resistance. Colonial domination has never been complete. Beyond the modern order—and in the middle of it—there have always lived multiple forms of existence that defy it. On this point, Deere draws above all on the work of Glissant to show that throughout the Americas, but perhaps with greater emphasis on the Caribbean, we can find forms of life that exceed colonial domination and decodify it. Glissant speaks of a world of archipelagos that makes the fiction of identity impossible. This is a world in which identity is rhizomatic and lacks a sole source. It is for this reason that the spatial patterns of modernity were incomplete in the Americas. They are always overwhelmed by other forms of knowledge, by other forms of inhabiting territory. When evaluated by the normative criterion of the European order, the Americas are necessarily seen as disordered and undisciplined spaces. In this same manner, if the Americas were the first laboratories of the modern order, they were also the first laboratories in which this order was openly defied.

The message of the book is clear: Modernity cannot be adequately thought without a theory of space that shows the darker side of colonization as much as it shows the local struggles for resisting and reconfiguring the nomos of capitalist globalization. It is not enough to highlight the hollowing out and destruction, as if the encounter between Europeans and non-Europeans could be uniquely reduced to domination. There were also processes of spatial transculturation (*transculturación espacial*) that permitted the creation of locales that defied colonial norms and in which Indigenous, Black, and European populations coexisted; border spaces that, indeed, could be seen as heterotopias in Foucault's sense. For this reason, the way Deere supplements the notion of heterotopia with that of heterarchy is interesting, implying that there existed—and exist still—spaces completely ungoverned by colonial hierarchies (of race, class, gender) in which it is possible to escape from the modern grids and inhabit territory otherwise. These are spaces that I would like to refer to, perhaps, as transmodern and that constitute the foundation to conceive a truly decolonial political philosophy.

I, thus, welcome the publication of this magnificent book and wish its author the greatest success. Deere's book is an example of the excellent work being done by a new generation of philosophers in the United States who are dedicated to reflecting on the colonial inheritances of modernity and their relation to Latin American thought.

Acknowledgments

It is hard to count the many sources of inspiration and support that went into the writing of this book. I do my best to share some of them here, with apologies in advance for the many omissions and noting that any shortcomings of the book are mine alone and not those of my esteemed friends, colleagues, and teachers.

Tracing my way back: I am grateful for my colleagues in the Philosophy Department and across the university at Texas A&M University who have supported my work for the last three years. Thank you, Andrzej Baginski, Alex Chinchilla, Benjamin Davis, Theodore George, Amir Jaima, Alberto Moreiras, Gregory Pappas, Omar Rivera, Adam Rosenthal, Kristi Sweet, and George Villanueva, among others, for making this a welcoming intellectual home. I also thank the graduate students at A&M for their great conversations, particularly my students from my graduate seminar The Spatial Turn, which developed on many of the themes in this book as it was nearing completion. I acknowledge the Melbern G. Glasscock Center for Humanities Research for generous support with a publication grant to help in finalizing the manuscript and also support in bringing Santiago Castro-Gómez to Texas A&M as a Glasscock Center Short-Term Visiting Fellow in Fall 2024. I thank José Alfredo Ortíz Angeles for his help with the final citations and bibliography and Lucas Scott Wright for his translation of Santiago Castro-Gómez's foreword to the book.

At Wesleyan University, I thank my amazing undergraduate students in my Decolonial Philosophy and Latin American and Caribbean Political Philosophy courses: many inspired me with contributions to our classes and a passion for imagining a more just world. Thanks especially to Ethan Barrett, Ivanie Cedeño, Manuel Domínguez, Mengmeng Gibbs, and Samela Pynas. Thanks also to colleagues and friends in Middletown: Stephen Angle, Carolina Diaz, Demetrius Eudell, Steven Horst, Joseph Rouse, and Elise Springer. At Fordham University,

I would like to thank Jeff Flynn, Crina Gschwandtner, Samir Haddad, and Shiloh Whitney, among others, for making it such a welcoming intellectual community and Aaron Pinnix for the great Glissant conversations. The graduate students and faculty in the Francophone reading group and Social and Political Workshop at Fordham offered wonderful spaces of intellectual engagement, often giving me ideas to reflect on for this book. At Loyola Marymount University, many thanks to my dear friends Andrew Dilts and Sina Kramer for all the conversations over happy hour and to Brad Stone for all his support. And thanks to my students, Nelson Peralta, Kalika Rudd, and Jake Hook.

At DePaul University, I can only begin to thank Elizabeth Millán for introducing me to Latin American philosophy as a field and her unwavering support along the way. Darrell Moore I thank for introducing me to the notion of "the ontological transformation of space" in his Early American Political Thought seminar. That idea, and so many others that we discussed over the years, have traveled with me through many stages of this project. María del Rosario Acosta López, Rick Lee, Michael Naas, and Kevin Thompson, among others, I thank for their support and critical engagements. Thanks to all my friends and colleagues from Chicago. Thank you to Alison Staudinger and Floyd Wright for friendship and care from Chicago all the way to Karaburun. Thanks to my teachers at Cornell University: Bruno Bosteels, Susan Buck-Morss, Michelle Kosch, Ken Roberts, Diane Rubenstein, and James T. Siegel.

Thanks to the intellectual communities that have supported and energized me through the last years in various venues, forums, institutions, and noninstitutions, many of whom are mentioned above; but I would like to add María del Rosario Acosta, Jimmy Centeno, Miguel Gualdrón, Dilek Huseyinzadegan, Eduardo Mendieta, Frederick Mills, Ofelia Schutte, Grant Silva, Alejandro Vallega, Ernesto Rosen Velásquez, Gabriella Veronelli, the Caribbean Philosophical Association, the APA Committee on Hispanics and Latinxs, Philosophies of Liberation Encuentro, and AFyL (Association of Philosophy and Liberation—United States, Mexico, and international).

Thanks to Bruno Bosteels and George Ciccariello-Maher for your interest in this project for the Radical Américas series at Duke and to my editor, Courtney Berger, for your support and patience all along the way. Thanks also to Laura Jaramillo and everyone at Duke University Press. I thank the two anonymous readers for their insightful commentary and suggestions on the draft of the manuscript. Their careful engagement has made this a better book.

I acknowledge the *Inter-American Journal of Philosophy*, where an earlier version of chapter 1 appeared as "Coloniality and Disciplinary Power: On Spatial Techniques of Ordering," in vol. 10, no. 2 (2019). I also acknowledge *Latin*

American Perspectives, where sections of chapter 3 are scheduled to appear as "Transmodern Geographies and Coloniality: On Enrique Dussel's Pluriversal Modernity" (March 2026).

I would like to thank the late Enrique Dussel for transforming my understanding of Latin American philosophy and shattering the possibility of a Eurocentric reading of modernity. His generosity and ferocity as a transformative thinker of Latin American philosophy has opened the field to many possibilities for doing philosophy in a transmodern key. I also thank Santiago Castro-Gómez for his support, friendship, and encouragement while I translated his book *Zero-Point Hubris* into English and, more recently, during his visit to Texas A&M as a visiting fellow. I also thank him for his generous engagement with this book in his foreword.

I could not have written this book without my partner, Ege Selin Islekel, who has been there since I first began forming these ideas nearly a decade ago. There are no words to describe what goes into the kind of encouragement and support she has given me all these years. Thank you for everything, Selin.

Many thanks to my mother and father, Don William and Claudia, for always encouraging me to pursue my dreams and for their love and support all along the way. Huge thanks also on this note to my sister, Cristina, and brother, Theodore. I thank my aunt Carmen Diana for inspiring me on the academic path and her profound example as a Latin Americanist. Por la sabiduría de la isla and to the memory of my *abuelita*, Carmen García Deere, I offer much gratitude. To the late Fernandito, to Tati, and to all my Puerto Rican family, thank you for the Caribbean inspiration. Türk aileme çok teşekkür etmek istiyorum. Many thanks also to all my large and lovely family from Colorado to Ohio to New Jersey.

Introduction

In the early sixteenth century, the Spanish empire made a map of its overseas territories called the Padrón Real. Kept in secret in the Casa de Contratación in Seville, all captains consulted the map prior to departure on transatlantic journeys and carried an official copy with them.[1] Constantly in development, integrating new additions whenever captains returned from voyages with new findings, this large-scale map charts an abstract epistemology of the globe from a patchwork of local experience of its navigators. This abstract knowledge is kept under lock and key, shared only with Spanish navigators and cartographers, a "secret science" that ties a new quest for global knowledge with the quest for global power.[2]

Jorge Luis Borges echoes the project of the Padrón Real in his story of an empire with the will to make a perfect map of its entire territory. The desire for exactitude and precision, the will to know every last microscopic detail of this territory and to grasp it on the abstract plane of the map's grid, consumed the technicians and cartographers of this fantastical empire to the point of absurdity. Borges explains, "The Art of Cartography attained such Perfection that the map of a single Province occupied the entirety of a City, and the map of the Empire, the entirety of a Province."[3] In time, the cartographers grew dissatisfied with the imperfections and insufficiencies of these maps and decided to make "a Map of the Empire whose size was that of the Empire, and which coincided point for point with it."[4] The next generation would, however, find this map to be of little use and leave it to deteriorate and decay in the sun and soil, leaving behind nothing more than tattered remnants of a once grandiose cartographic project.

With this story, Borges exposes not only the absurdity of a will to perfect representation but also that of an imperial will to know and to order, marking a desire to grasp every last point of the globe with precision and accuracy. Yet when this project is taken to its logical breaking point, it cannot hold together.

Indeed, it cannot hold the disparate moments and points of the territory together without attempting to replicate the very thing it wishes to grasp. Representation is not enough unless it reproduces the original, revealed in the paradoxical reversion to the materiality of the map; the absurdity of making a map that must materially extend out across the entirety of the territory and match its points of reference one by one. The dream of empire is, in short, the invention of a perfect order that would be not only mapped and known but also shaped and controlled at the material level. Beyond the representation of order, material order must itself be invented, produced.

The total map, the project of creating a perfect mode of representation, cannot find an adequate site to ground its order, as seen in the absurd attempt to match the map to a one-to-one scale. Knowledge requires a site where words and things come into a possible space of relation. As Michel Foucault shows in *The Order of Things*, without this meeting point, there is no order of knowledge or science.[5] For modern knowledge, the abstract grid is the ordered site where the classification of things in the world is made possible. Borges shows the impossibility of that ordered (cartographic) grid, its inability to serve as a solid foundation for knowledge or power. Yet the order of the grid is not just cartographic. It is more generally epistemological (abstract tables of classification of plants, animals, minerals, or human races found in the encyclopedic desire to classify everything) and also political-material, the organization of human bodies in space such that they can be studied and controlled in the urban space of the colonial city, the plantation, the school, factory, hospital, or prison.

The Padrón Real reveals a patchwork approach to the early dimensions of this ordering project, as individual pieces of experience are glued together on the abstract grid. The exploring subjectivity of the navigator's experience is charted out by the abstracting work of the cosmographer and cartographer. The local experience of space and territory brought into the order of the grid, to empty experience of its empirical content and specificity, fitting it onto a flat coordinate plane.

There is a tension that should not be overlooked between this imperial will to know that empties space of its local specificity and the establishment of a new mode of ordering space. The Americas operate as the laboratory for this double project of order: a project that begins as an imperial will to know and a violent will to subject the "other" to fit onto this grid. The project is not only violent and destructive but also productive of new regimes of space and subjectivity, where the violence of emptying is paired with the productive discipline of ordering. The ordering of this projected emptiness is epistemological and political as it shapes subjectivity, knowledge, race, bodies, and daily habitus in the Americas and across the Atlantic triangle. The project of fitting onto the grid has to do not only with space and geography but also with the ordering of humans, plants, language, and ideas into an organized system of knowledge and power.

Borges further illuminates the absurd limits of ordering projects in his famous account "The Analytical Language of John Wilkins," on the creation of a universal language by the seventeenth-century British philosopher John Wilkins, modeled on a metric system in which "each word defines itself."[6] Wilkins's language parallels the project of the total map in that a perfect system of reference and order might be developed such that nothing would fall outside it. The system of representation establishes its own perfect order, no longer dependent on the original.[7]

The ground of order for Wilkins's project is upended when Borges refers to an apocryphal Chinese encyclopedia that categorizes animals in divisions, such as "a) those that belong to the emperor . . . g) stray dogs . . . h) those that are included in this classification . . . k) those drawn with a very fine camel's-hair brush . . . n) those that at a distance resemble flies."[8] Foucault describes the shattering force of this monstrous classification system in his preface to *The Order of Things*. Borges's encyclopedia inspires Foucault's project as a jolt of lightning that wakes you from a slumber. It provoked "laughter that shattered . . . all the familiar *landmarks* of my thought—*our thought*, the thought that bears the stamp of our age and *our geography*—breaking up all the *ordered surfaces* with which we are accustomed to tame the *wild profusion of existing things*."[9] The apocryphal encyclopedia that broke up the ordered surfaces of his thought contains a system of classification for objects that is absurd, arbitrary, and groundless: a categorization of animals with no other principle of organization than the alphabetic enumeration of a list.

Borges's work brilliantly exposes the breaking point of projects of totalization, as seen in both the preceding stories. He shows the "wild profusion of things" underneath the taming of ordered grids. He illuminates the groundless ground of classification. Borges's project reveals the limits of totalizing classification, in general; however, I consider what this kind of story has to say about the invention of order in the Americas, in particular. How does the history of the Americas after 1492 evidence a certain obsession with ordering space and knowledge?[10]

The heterotopia of Borges's thought, its troubling obsession with the breaking point of order, demonstrates something about the space of the Americas and its relationship with European space.[11] Indeed, if we consider that the commencement of a new project of ordering begins with the European conquest of the Americas, the heterotopia of Borges's writing is not about an orientalist fantasy of Chinese culture but about an interest in the ordering of the Americas, a critique of the colonial project of totalization. Foucault writes (about the East, but rereading this passage here with reference to the Americas), "There would appear to be, then, at the other extremity of the earth we inhabit, a culture entirely devoted to the ordering of space."[12] In fact, this devotion to the ordering of space is a project born of the colonial struggle of modernity, as Europe works to impose a grid on the Americas and the globe. It is a question not of a completely other culture but of the mirror of Europe itself in its colonial entanglements: the Americas as the very intensive site, the heterotopic space in which the project of order emerges. This ordering project of the Americas will also have a boomerang effect in defining the shape of modernity in Europe.

Borges's obscure encyclopedic references to the history of Western thought often distract his readers from this history of (Latin) American space, the emplacement of his own thought. I situate his atlas of the impossible, instead, as definitive of a radical and critical thought of the Americas. In this sense, Borges takes the tools of Western philosophy and subverts them in a critical gesture, a gesture that is definitive of much Latin American philosophy and, more recently, decolonial thought. As he writes, "To have appropriated their weapon and turned it against them must have afforded him a bellicose pleasure."[13] A critical thinking of global modernity takes stock and engages the cartography of this battlefield subversively.

The Invention of Order

The invention of order refers both to an abstract table of classification in which words and things meet and to the material table in which human subjects and material spaces are ordered and organized across the globe. Invention emerges out of the colonization of the Americas, as the commencement of a global problematic of space. New spatial concepts and practices emerge from the wreckage of old European, Indigenous, and African worldviews. If we can use the phrase *New World*, it is precisely to refer to a new reality, beyond these previously existing separate worlds of Indigenous Abya Yala, Africa, and Europe.[14]

The notion of empty space is at the heart of these spatial transformations. A notion invented by European practices and sensibilities in justifying colonial conquest, empty space is practiced by depriving existing Indigenous, African, and mestizo populations of their spatial distributions and rights to land. The prehistory to this notion is written in the shift from a prior notion of uninhabitable space attributed to the margins and extremes of the unknown world. As Sylvia Wynter shows, prior to Columbus, Cape Bojador on the western coast of Africa is seen by Europeans as the *nec plus ultra*, the limit of all habitable space on earth.[15] Shortly after Columbus travels to America, the transition is made from seeing this space as monstrously uninhabitable to potentially habitable (empty) and, thus, in need of order.

In this respect, the medieval worldview shifts in the late fifteenth century, which can be seen in the papal bulls of this period and the imperial designs of the Catholic monarchs. As early as 1493, global lines are drawn by these European powers: On one side they see their own space as organized and accounted for; on the other side, empty space, free for exploration, discovery, and appropriation. On May 4, 1493, Pope Alexander VI marks out such a meridian line to "donate" the land beyond for discovery and settlement by the Spanish. In 1494, the Treaty of Tordesillas draws another line, on the meridian 370 leagues west of the Cape Verde Islands, to settle claims between Portugal and Spain.[16]

The shift in European worldview from uninhabitable to empty space ready for order marks this initiation of a new global problematic of space. This problematic is not limited, however, to the drawing of abyssal lines that mark out supposedly empty space from organized, civilized spaces.[17] The other side of the line is defined by not only ontological negation and emptiness but also the production of a whole regime of order, connected to disciplinary and racializing practices of shaping Indigenous, African, mestizo, and criollo subjectivity. A regime of order that is also at the heart of the ordering of modern systems of knowledge.

The global problematic of space is a central theme of decolonial thought and critiques of coloniality. For example, Enrique Dussel's work has been key in tracing the geopolitics of knowledge with its global lines that divide between center and periphery, totality and exteriority.[18] In his work, he emphasizes that beyond the line, in the zone of exteriority, is the space of nonbeing. The European center construes itself as the space of being, while it negates the being of the other in the periphery. The history of modernity, for Dussel, is also the history of the ontological nihilation of the periphery that began with 1492. Philosophy of liberation is, what Dussel terms a barbarian philosophy, an affirmation of the periphery and the creativity that surges forth from beyond the domination and determinations of the center. To be clear, then, philosophy of liberation argues for the metaphysical *reality* of the other who is beyond *being*. Justice, creativity, freedom, and the other are all exterior to the determinations of the dominant system of being. I propose, however, to read Dussel's concept of the other in a material sense, as the subject who is materially excluded and silenced by practices of power. In this sense, I look at the materiality of the *production of the other* as subject.[19] Coloniality and its spatial regimes are thus both nihilating and productive regimes of power. There is a tension between these two dimensions that must be traced.

Dussel's writings are foundational for the work that has developed around the concept of coloniality. Coloniality suggests that the structures of power and knowledge (particularly consolidated around race and gender as they articulate a hierarchy of labor[20]) that emerged with the history of colonialism continue to fundamentally shape the supposedly postcolonial world. Coloniality also shows that modernity was never separate from its colonial history. Thus, the suturing together of modernity/coloniality is a corrective decolonial concept to show that modernity is not free from this vector of coloniality.[21] The modern project to illuminate the world through reason is entangled with a history of colonial violence and ordering. Colonial violence engages in a double movement of emptying space while also producing ordered spaces. To understand modernity/coloniality we must understand the production of this regime of order.

Boaventura de Sousa Santos coined the term "abyssal thinking" to refer to the epistemology that emerges with this ontological nihilation of the periphery and what is beyond the line.[22] Dussel and Santos offer an account of what is beyond the line and the negation of the other, while I argue that a richer understanding of the production of order is needed to understand the dynamic of space on the other side of the line. The local production of colonial subjects and spaces will also be crucial to understand modes of resistance and creativity that escape or counter coloniality. While the Spanish conquistadors often construed "beyond the line" in terms of ontological negation, the production of coloniality, in fact, also involved a complex engagement and ordering of these spaces. Undoubtedly, these two dimensions of coloniality are entangled, as the nihilation and emptying of the periphery will serve as a condition of possibility for the ordering and production of space.[23]

In this sense, I offer a productive and not simply nihilating reading of coloniality. This approach also allows for what Santiago Castro-Gómez has termed a "heterarchic" reading of coloniality. He argues that the theory of coloniality as a global theory has placed its analysis primarily at the macro level.[24] Aníbal Quijano, Dussel, and Immanuel Wallerstein develop this global account of power relations.[25] Under this model, coloniality is thought of as a global system of power that is inextricable from capitalism. Furthermore, the analysis of coloniality remains at the level of a macro narrative of the emergence of this new global structure. Castro-Gómez points instead to a *heterarchic* conception of power (in the sense that there is not one but multiple foundations, or archai, as opposed to a top-down *hierarchical* model[26]), which includes the global level of coloniality but also accounts for local and regional practices of power. No one level of power is strictly determinative, but each is influential for the global level and vice versa while not being reducible to the other.[27] The hierarchic conception of power, in contrast, would be from the top down, and the macro level would determine all other micro and meso levels. The heterarchic, instead, sees the production of power relations to be an open-ended process in which different levels may intersect at different moments in different ways. Throughout this book, I work between these local, regional, and global levels, employing a heterarchic method to understand the coloniality of space and the invention of order as a practice of power, knowledge, subjectivity, and racial formation. The heterarchic reading also opens the possibility of thinking resistance to coloniality at the local and regional levels. On this point, I develop on María Lugones's understanding of resistance in terms of practices that do not operate as a pure outside to power but rather as immanent to everyday practices within oppressive regimes of power and coloniality. Lugones opens the possibility of constructing other resistant worlds within and to the side of these oppressive regimes.[28]

Accounting for Modernity and Space

This book argues for a spatial reading of modernity, thus, highlighting the colonial and global dimensions of modernity. I argue that modern thought is forged through global emptying and ordering of space. Modernity takes place in a global battlefield of space that seeks to neutralize and eliminate other modes of spatialization while imposing and producing a single unitopic model of space.[29] The grid neutralizes, empties, and controls other spaces. It is the heterotopia that ultimately aims to create the globe as unitopia.

Space is the framing problematic of this book. My reading of space is fundamentally influenced by Immanuel Kant's aesthetic in *The Critique of Pure Reason*, where he argues that the conditions of possibility of all experience are grounded within space and time.[30] No experience can take place unless inside these spatiotemporal conditions. Modern thought and philosophical accounts of modernity, however, have prioritized time over space. Theories of the subject turn to questions of time, memory, and consciousness in a Cartesian void of space. Theories of modernity rely on notions of progress and emancipation from the past of tradition without asking where these temporal transformations took place, assuming that they emerged in complete isolation from the space of the globe (or moved in a linear fashion from Greece to Rome to Germany and France). The philosophy of modernity wishes not to be tied down to a space of conflict, to take place in a neutral container that happens to coincide with and only with the space of Europe. As Kant argues in *The Critique of Pure Reason*, the powers of reason pretend not to be conditioned by their aesthetic conditions of possibility. He draws an analogy to a dove in free flight that wishes to escape the resistance of gravity, as if space could be airless, empty, and frictionless.[31] Philosophers of modernity have followed this dove in search of an airless space without resistance, a space so smooth and frictionless that it no longer conditions thought at all. Instead, my argument is that modernity is constituted *within* and *as* a space of struggle. Unitopic space seeks totalization, but the story does not end there: Space has a history of resistance, formation, production, and transformation.

My account intervenes against the standard accounts of modernity represented most prominently by thinkers such as Georg Wilhelm Friedrich Hegel, Max Weber, and Jürgen Habermas.[32] These thinkers either forget the spatial nature of modern concepts or seek to defrictionalize space into an empty container. Even Hegel, the great dialectician, reduces space to the unitopia of the Mediterranean: All history condenses around and flows toward one center. Space is either empty and without friction or else it is hierarchized by a unidirectional movement of history, with all progress flowing from East to West toward the Mediterranean. In this Eurocentric spatial account, Europe is supposed to be the unique and neutral site where universality emerges and with that all global entanglements with colonialism magically disappear.

Thus, these traditional accounts fail to think the map of modernity, the cartography from which it emerged. Their cartography is Eurocentric while also pretending that space is neutral and empty, just a container. Modernity is, in this sense, inscribed as a temporal concept by its very name: To be modern is to come after, to have followed a line of exclusively (or ultimately) European development and progress. By the eighteenth century, the spatial dimensions of modernity will be embedded in temporal terms of progress: maturity versus immaturity or modern versus primitive. This is a temporalization that forgets the spatial battlefield that constitutes the site of such a division. It both forgets the role of the periphery in the constitution of Europe as center and temporalizes the periphery as traditional, backward, in the past, savage, or not yet modern. Space is hierarchized and subordinated to time. Space is temporalized as in the present cutting edge of progress or in the past of tradition and immaturity.[33]

To understand the practice and organization of modernity, we need to analyze the production and ordering of space.[34] The importance of thinking space, modernity, and coloniality together has been highlighted by decolonial thinkers from Dussel to Lugones, to Mignolo, and Wynter.[35] Critical readings of space (generally with little attention to coloniality and the globe) have also been key dimensions of European critiques of modernity from Foucault to Gilles Deleuze and Felix Guattari, to Theodor Adorno and Henri Lefebvre. Foucault, for example, points out that space is the forgotten field of analysis of modernity, while time is always privileged in its stead. The richness of subjectivity and its temporality, especially in the phenomenological tradition, has been analyzed with great depth, while space is thought of as an empty and immobile container. "Space,"

Foucault writes, "was treated as the dead, the fixed, the undialectical, the immobile. Time, on the contrary, was richness, fecundity, life, dialectic."[36] My analysis aims to avoid the retreat into the frictionless, undialectical, and immobile space: a space that is additionally temporalized. Modernity has been prone to a specific mode of forgetting: the forgetting of space, its dynamics, its methods of production and ordering, and its role in the constitution of the human subject and the ordering of knowledge and power, a history of production and ordering that is additionally colonial and global at its core.

While Kant's aesthetic suggests that space must not be forgotten in the constitution of human experience, he still ultimately sees space in this fixed, undialectical mode. Foucault's reworking of Kant shows us that the conditions of possibility of experience of time and space are not transhistorically fixed but are shaped in relation to different epistemes and regimes of power. His work offers resources to think about the production and ordering of space in the construction of new epistemes and new modes of power relations. Yet Foucault neglects to account for the global and colonial practices that are so central to modern spatial orders.[37] This book brings the resources of Continental European thinkers of space like Foucault into dialogue with thinkers of space in the Americas such as Enrique Dussel, María Lugones, Sylvia Wynter, Santiago Castro-Gómez, and Édouard Glissant. On one hand, critical readings of spatial modernity from European thinkers offer rich resources, but they are made to address a new set of problems when they travel beyond the shores of Europe. On the other hand, these accounts are supplemented and critiqued by the methods and insights of decolonial thought in the Americas, from the United States to Latin America and the Caribbean. Decolonial accounts have situated space as a central problematic of colonial modernity, yet a more detailed account of the interplay between the emptying and ordering of space in the modern period is needed, what I describe as the coloniality of space.

The notion of the coloniality of space is also tied to a tradition of Latin American thinkers who began developing an understanding of the relationship between space, colonization, and modernity starting in the middle of the last century with the Mexican historian Edmundo O'Gorman's 1958 book *The Invention of America*. The importance of O'Gorman's work is perhaps rivaled only by the 1984 book of Ángel Rama, *The Lettered City*, which rethinks the relationship between space, writing, and power in the Latin American city from the colonial era through the twentieth century.[38] While O'Gorman is perhaps the first to explore this epochal historical shift in ontological (the colonization of being) and spatial terms (America as the *invention* of a new understanding of being), Rama offers an account of the spatial-ordering principles of Latin American colonial cities and how they relate to the organization of an episteme of knowledge in Foucault's sense via writing. I draw on this notion especially in chapter 1, centered on the question of order.

Chapter Descriptions

This book is divided into two parts, each consisting of two chapters. Part I, "Genealogies of Colonial Space," develops a historico-genealogical account of the ordering of space in the Americas from the long sixteenth century through the nineteenth-century nation-building period.[39] Chapter 1, "Orders of the Grid," explores the heterotopic implantation of the grid in the Americas through the ordering of urban colonial space. Sixteenth-century Spanish colonial cities are one of the first models in the newly global world of an extended project of gridded and plotted urban space. I argue that the grid develops a new role in shaping human space and takes on a protodisciplinary and racializing role to make colonized bodies productive and docile while increasing the extraction of resources from the Americas. This grid space is constructed as a heterotopia, an other space outside the given spaces existing in Europe. In this way, Europe invents a new mode of order, previously unthought in Europe, in the Americas. I consider also how this gridding impetus to order space extends beyond the colonial city to the whole countryside during the nation-building projects of the nineteenth century. Here I develop on the racial ordering of space next to the racial anxieties of disordered and mixed space that lies beyond the control of the grid.

Chapter 2, "Orders of Movement: The Traveler and the Settler," analyzes how the Americas were conceptualized and produced as an empty space free for the appropriation and ordering of European projects. I show how this conception of emptiness is coupled with a codification of certain spaces as free for the movement and settlement of certain subjects: the racialization of space. The traveling subject with the right to move about is given expression in Francisco de Vitoria's 1532 text, "De Indis." Yet I ask whether such a notion of mobility would be reciprocally applied to the Amerindian traveling to the shores of Europe. I show that the Janus face of the traveling colonial subject is the settler subject, who seeks to plant roots and distribute a new order of space. Turning to John Locke's theory of property and its notorious connections to North American Puritan settler colonialism, I show how other modes of distributing and inhabiting space are excluded, particularly Indigenous ones, in this settler model. Thus, motion and settlement are not opposed concepts but instead racially codified around certain modes of subjectivity of who can move and who can settle, and what counts as legitimate movement and legitimate settlement. The Indigenous notion of refusal to state-based recognition projects can be read in this light as a response to Vitoria. I turn to María Lugones's radical understanding of space, in conclusion, to open questions of resistance to the coloniality of space in everyday local practices.

Part II, "Transmodern Cartographies," draws on contemporary reflections about spaces of resistance in the past and future of modernity in Latin American and Caribbean thought. Chapter 3, "Transmodernity and the Battlefield of Coloniality," develops a spatial reading of Enrique Dussel's history of modernity alongside his theory of transmodernity. The material invention, production, and silencing of the other, the non-European, is a struggle that is at the roots of European claims to universality in modernity. Dussel finds that this problematic is embodied in the figure of the *ego conquiro* (I conquer) emblematized by the conquering subjectivity of Hernán Cortés and the 1519 conquest of Mexico. The I-conquer figure forms the prehistory of Descartes's *ego cogito* (I think), in which Descartes's epistemology is built on the spaceless and frictionless ground of the zero point; a spaceless ground predicated on the forgotten history of a dominating subjectivity.

Chapter 3 traces the suggestion that embracing a truly global conception of modernity requires a pluriversal notion of reason that would overcome the violent excesses of the Eurocentric modernity, what Dussel refers to as transmodernity. I analyze Dussel's notion of the global silencing of the periphery in the birth of modern European reason in parallel to Foucault's account of the silencing and spatial exclusion of the mad and the poor across Europe in the latter's account of the birth of modern reason. I conclude the chapter with the question of how to break open the universalizing position of knowledge production to the pluriversality and plurality of epistemological spaces of exclusion and subjugated knowledges of modernity without forgetting or reifying the violence of the battlefield of modernity.

Chapter 4, "Archipelagoes of Resistance: Limits of the Map," develops on these transmodern questions by bringing a resistant understanding of the Caribbean archipelago into dialogue with coloniality of space in the Americas more generally. Through Édouard Glissant's account of a novel Caribbean geopoetics built on the history of destruction and uprooting, I consider practices of resistance and an aesthetic imaginary of resistance to the modern global project of order. This final chapter develops the turn from an account of spatial forms of domination in the Americas to possibilities of creative resistance and alternative modernities that are not simply condemned by violent histories. Glissant's archipelagic spatial model of the Caribbean is a network of relations offering an alternative vision of modernity, against the totalizing colonial rootstock that imposes one topos of space. Returning to Lugones's account of resistance alongside

Glissant, and the Zapatista Indigenous movement, I argue that the coloniality of space is never complete in its impulse to empty and order. There is an excessive, open, and irruptive landscape that these movements draw from.

In bringing Afro-Caribbean thought in dialogue with Indigenous, Latin American, and Latinx thought more generally, we find the tense meeting point between these histories and geographies to consider how the Afro-Caribbean and Latin America share entangled histories in the coloniality of space while also demonstrating unique forms of spatial domination and irreducibility between the Caribbean plantation and the urban ordering of Spanish American space. I turn to Glissant's notion of the irruptive and open landscape of the Caribbean islands with his invented term *irrué*. This notion serves as a hinge to think about what is excessive and not captured by the ordering impulse in the landscape of the Caribbean and the Americas.

Glissant's aesthetic, relational, and site-specific understanding of creation and thought furthers my global account of transmodernity, to draw on not only a pluriversal notion of reason but also creative practice more generally. I bring Glissant into dialogue with Dussel to further develop a spatially situated account of modernity articulated across a network of relations and locations. Dussel's account of transmodernity, affirming the reason of the other, is helpfully supplemented by Glissant's aesthetic view of reason and relation. Relation thought in terms of the affective, aesthetic, and imaginary, which does not exclusively privilege the rational, opens onto a richly decolonial vision of transmodernity. I turn also to the Zapatista Indigenous rebellion in Chiapas as an example of transmodern movement whose word has echoed across languages and soils and formed horizontal relations that point to another kind of spatial distribution of the world that fits many worlds, beyond the invention of order.

PART I. Genealogies of Colonial Space

1. Orders of the Grid

The evil that besets the Argentine Republic is the expanse [*la extensión*].
—D. F. SARMIENTO, *Facundo*

On August 2, 1513, King Ferdinand II writes the following royal directive to Pedrarías Dávila, the conquistador and future governor of Panama: "Let the city lots be ordered from the start, so once they are marked out the town will appear well ordered as to the place which is left for a plaza, the site for the church, and the sequence of the streets; for in places newly established proper order can be given from the start, and will remain ordered without work or cost [porque en los lugares que de nuevo se hacen dando la orden en el comienzo, sin ningún trabajo ni costa quedan ordenados]."[1] Spanish colonial cities of the Americas are built without the walls and defensive enclosure defining medieval architecture in Europe.[2] Instead, they are constructed according to principles of order and the grid. In these cities, spiritual walls replace the heavy fortification of brick and stone.[3]

As the grid is deployed over and above the wall, a new relationship between human subjectivity, the body, and space emerges in the long sixteenth century. Beyond these localized urban colonial spaces, the Americas and the Atlantic at large became spaces of a new global history: The globe emerges for the first time as a material object to be circumnavigated, mapped, and known (on the grid). Both local gridding of cities and global gridding of the earth through cosmography and cartography emerge as central colonial concerns. This new spatial ordering is what I refer to as an *invention* in the sense that it emerges through new techniques of power and knowledge that organize human communities in new ways while also grasping the globe as a whole. This invention transformed the meaning, practice, and construction of space and spatiality.

In particular, this chapter considers how a new technique of ordering and producing space emerges in the sixteenth century, when the Americas were conceived as a heterotopic laboratory for the space of the grid. The Americas operate a compensatory function for the messy space of Europe as they set out to establish a new regime of order.[4]

Spatially, coloniality involves a global matrix of power. It grasps the globe on the abstract coordinates of the map, but it also involves the local production of order on the ground, thus connecting with formations of disciplinary power. The grid is produced at both local and global levels and not simply imposed from the top down, what Santiago Castro-Gómez refers to as the heterarchic logic of coloniality.[5]

The invention of ordered space in the Americas refers to the invention of another mode of organizing space, treating the Americas as a laboratory of empty space. This laboratory-like practice gave rise to the production of new techniques of order and protodisciplinary modes of subjection. The ordered gridiron space lightened the physical fortification of heavy walls and aimed to implant new methods of ordering the behavior of the human body and soul.

In tracing out this question of the heterotopic laboratory, I turn to a close reading of the disciplinary and spatial dimensions of the 1512 Spanish legal document aimed to codify colonial behavior in the Americas, the Laws of Burgos, followed by an analysis of the 1573 Royal Ordinances on City Planning issued by the Crown. These latter ordinances demonstrate the formalization of the spatial technique of the grid and disciplinary rules for the construction of the space of the city. I also consider the intersections between sovereign and disciplinary spaces that are at work in the sixteenth century in the interchange between colony and metropole.

Finally, I consider how the spatial ordering of the Americas gives rise to racial and moral geography. The ordering of subjects and spaces entails the spatial construction of race and the mapping of populations. I focus particularly on the spatial consolidation of the Indian in the early sixteenth century as a racialized subject in the ordered town of the congregation.[6] The emptying and gridding of space involves not only destruction and nihilation of the other but also the racialized production of the other. As Daniel Nemser argues with respect to colonial space in Mexico, race is spatialized and space is racialized, especially in the colonial context.[7] Moving from the account of order in the sixteenth-century congregation of the Amerindian to later accounts of moral and racial geography in the eighteenth and nineteenth centuries, I consider how the racial logics of colonial space carry over into the late colonial and early independence periods in Latin America.

The disordered countryside becomes an image of barbarism and a threat to the civilizing project of order. The expansive geography of the pampa in Argentina, for example, will be situated as a place of moral degeneracy and disorder, tied to the dangers of race mixing. The gauchos of the Argentine countryside are considered doubly degenerate, because their geographic circumstance and tendency to race mixing lead to laziness and moral backwardness. Thus, we consider the threat of disorder and its racial anxieties that lurk on the other side of the production of order.

Heterotopias of the Grid

The grid emerges as an apparatus of ordering across various practices and institutions in related, interlocking, and sometimes, discontinuous fashions.[8] The (re) birth of the grid in the sixteenth century organizes space at the practical and epistemological levels in early modernity/coloniality, in such a way and to such a degree that was unthinkable in previous centuries.[9] The grid refers here not only to the spatial mapping of geographic locations onto a coordinate system but also to a network of power and epistemology operative for the organization of language, writing, and urban space.[10] The grid is a matrix for the ordering of geographic locations and physical bodies into and onto a coordinate system that can be known and controlled, and it is an epistemological practice for the ordering of knowledge onto an abstract plane. In this sense, the grid serves as a fortification for the ordering of human behavior that does not rely on the restrictive physicality of walls but instead on the geometry of sight lines and ordering of bodies in space. Before the factory or the panopticon, there was the colonial grid.

Colonial techniques of organizing space emerge as a heterotopic laboratory for the organization, flow, and order of communities within a space. The problematization of space in the European scramble to respond to the opening of what they blindly consider empty space *releases* the technology of the grid and a new exercise of governmental power. Thus, the ordering of gridded space is the correlate of this notion of emptiness. Spain did not create gridded space ex nihilo; rather they implemented and exercised it in such a way that it gave birth to a new reality and a new political epistemology. The invention of the grid refers, thus, to the emergence of a new political apparatus, whose forces were released through a heterotopic practice.[11]

A heterotopia is distinct from the nonspace or ideal space of a *u-topia*, literally meaning no space or good space. A *hetero-topia* is, instead, an *other space*, a space that is constructed and organized according to a set of rules different from those that organize and constrain the rest of a society.[12] Heterotopias can involve, on one hand, sites and methods of resistance: the maroonage (*cimarronaje*) of escaped-slave communities, the flaneur who carves out new paths in the monotonous city, or the construction of communes as alternative modes of social and spatial organization. On the other hand, they can be laboratories for the construction of new forms of power and subjugation: the prison, the school, the factory, spaces of confinement, the plantation, or the colony. Or, beyond the binary, they may simply be places where normal rules of space and time do not apply: airports, train stations, resort towns, ships at sea.

While Foucault refers to the sea as a vast reservoir of heterotopias, we might further specify the Americas, conceived as empty space beyond the line, as the ultimate heterotopic reservoir for the construction and production of new modes of order in the early modern European project. The emptying of American space is clear from the 1493 papal bull and 1494 Treaty of Tordesillas that draw global lines beyond which all space is free for Spanish and Portuguese appropriation. The epistemological and political violence of this blanket emptying is incredibly destructive, yet if we continue beyond the line, we see that emptying is coupled with ordering.

In the sixteenth century, many of these ordering projects beyond the line are inspired by utopian principles of creating a perfect community in this "new" space. Given their materiality and their compensatory function with respect to existing spaces in Europe, these utopian aspirations are in fact embodied as heterotopias. Some heterotopias follow this utopian aspiration to resolve and overcome all existing contradictions within the normal space of society, through the perfect ordering of an *other* space, leaving the tangled space of Europe behind. There are many examples of these practices unfolding in the Americas in the sixteenth century and on, especially among Dominican and Jesuit religious orders. The Jesuits in Paraguay who attempted to make perfect communities and modes of life with the Tupí-Guaraní people is exemplary;[13] as are the attempts by Vasco de Quiroga, the first bishop of Michoacán, to implement principles from Thomas More's *Utopia* in Indigenous communities in sixteenth-century Mexico.[14] These cases push us to consider the ways in which domination is not always a relation of violent submission but is also one that seeks to produce new modes of subjectivity. The colony as heterotopia creates relations of domination through its impulse for ordered space and subjectivity, also leading to complex practices of resistance.[15] The colonies function as laboratories for the distribution and organization of human spaces beyond the shores of Europe, outside the spaces given within their society. This is especially clear in the utopia-oriented heterotopias that sought to create a perfect Christian pastorate and regime of evangelization and care.

In Ángel Rama's famous account of writing and space in the Americas, *The Lettered City*, he explores this compensatory dimension of American heterotopias: "Over the course of the sixteenth century, the Spanish conquerors became aware of having left behind the distribution of space and the way of life characteristic of the medieval Iberian cities—'organic,' rather than 'ordered'—where they had been born and raised."[16] The techniques of empty space in the New World made possible a new relationship to urban space, a planned order different from the organic development of medieval cities. That is to say, medieval European cities were usually built up in an organic fashion, spiraling outward with jagged streets like the arrondissements of Paris, rather than preplanned according to an ordered gridiron structure. In this sense, it is worth remembering that colonial spaces are often overlooked as derivative and instrumental to more fundamental cultural and architectural expressions in the metropole. Instead, this relation should be read in reverse. The sixteenth-century transformation of space occurs with its greatest intensity in the colony and not in the metropole: This is the heterotopic logic of coloniality, its boomerang effect.[17]

Foucault's account of disciplinary power operates according to a similar heterotopic logic, yet his account is limited to spaces within Europe. Disciplinary techniques of power emerge precisely through a construction of other spaces outside the normal confines of society: factories, prisons, schools, hospitals, and military camps. These institutions construct their own rules of spatial distribution and organize the space and time of the body in fundamentally unprecedented ways, giving birth to a new apparatus of power. The construction of a docile body is precisely predicated on this production of a new spatiotemporal order in which the subject is enmeshed within a school, a factory, a prison, or a hospital.

Despite his reliance on European examples, Foucault indicates that disciplinary power could just as easily be read as emerging out of practices of colonialism and slavery,[18] two topics about which he writes little in his corpus.[19] In a brief footnote in *Discipline and Punish*, he writes, "I shall choose examples from military, medical, educational and industrial institutions. Other examples might have been taken from colonization, slavery and child rearing."[20] It is worth asking how following out the genealogical insights from the space of the colonies would shift Foucault's characterization of disciplinary power. While Foucault only hints at this in his footnote, there is an enormous archive to unpack and consider how techniques of power tied to coloniality might shift our understanding of disciplinary power and its European provenance.

This leads us to ask whether colonialism merely involves or itself invents disciplinary institutions and apparatuses. The question is whether sixteenth-century Spanish colonialism is a disciplinary practice and if we can identify the birth of the grid as a moment in the spatial construction of the disciplines. Foucault identifies the birth of disciplinary practices with the shift from Renaissance notions of ideal types and stable natures (the ideal sixteenth-century soldier with their discernible marks of courage and nobility) to the seventeenth and eighteenth centuries when a new politics of the body was born, an anatomo-politics. The notion was that a body could be molded to become more forceful and useful as it also became proportionally more disciplined and obedient: the increase of extraction of forces alongside the increase of subjugation. Foucault writes, "Discipline increases the forces of the body (in economic terms of utility) and diminishes these same forces (in political terms of obedience)."[21] Given Foucault's periodization focusing on the seventeenth and eighteenth centuries, it might be surprising to find disciplinary institutions emerging in sixteenth-century colonies. However, if we follow the definition of discipline as the increase in extraction of forces from the body in proportion to increases of political obedience, we are faced squarely with a fundamental dimension of both colonization and slavery. Furthermore, if we consider how discipline produces a new mode of self-policing subjectivity, we also find these elements in the project of Christian evangelization of Amerindians. The intersection of coloniality and discipline points us also to the anatomo-politics of the construction of race in the colonies.

As Aníbal Quijano has argued, the birth of European colonialism in the long sixteenth century gave rise to new structures of the domination of the body in terms of race and labor, a matrix of power and knowledge that he refers to as *coloniality*, which continues to exert force more than five hundred years later despite the end of most direct forms of colonial rule.[22]

Colonization and slavery give rise to power structures that are concerned with extracting forces from the body while operating a wholesale domination and restriction of the political forces of the body in terms of obedience. Race served as a justificatory principle to hierarchize different practices of labor exploitation and subjugation. One principal question of difference to be raised with respect to the anatomo-politics and the disciplines of seventeenth- and eighteenth-century Europe versus those of the colonies and the plantations is the way in which force is extracted from the body and the way in which domination is enacted. Violent forms of submission appear quite differently from subtle tactics of coercion.[23] The pastoral techniques of heterotopias, constructing new quasi-utopian religious communities, surely involved a double-edged deployment of violence and care, as Daniel Nemser argues.[24] Violent destruction of previous ways of life and habitats for the production of the new. In this process, the intensive ordering of these spaces through pastoral care also leads to the production of the racial category of the Indian. That is to say, the consolidation of many heterogeneous ethnic and linguistic Native groups into one homogeneous group that can be gathered together in space, ordered, and made to incorporate a new habitus. The Indian did not exist prior to 1492; instead this racial category is created through spatial practices of ordering.

Race in this sense is not only about justifying violence against the Indian but also about justifying a new regime of power. Thus, the colony as heterotopia is involved not only in violent submission (and ontological negation) but also in the production of new forms of subjectivity through the construction of gridded space alongside the inculcation of Christian morality. We find violence paired with pastoral care. The presence of care in no way signals the absence of violence but rather points to how coloniality produces racialized techniques of extracting labor, organizing the family, and ordering bodies in space.

At the same time, one might suspect that violent forms of domination occur only under models of sovereign power, a form of power that tortures the body and marks its signs of superior force directly on the body. However, colonization and slavery also give rise to a microphysics of power relations around the body. Their aim is not just to mark the body, torture it, destroy it, or stun it into submission: The colonial project has as a primary aim the extraction of forces from the body and doctrinal conversion of souls. To achieve this increase in force and conversion, the colony develops a microphysics of organizing all the minute details of how a body is used and exerted throughout a day, a year, a life. To organize these minute details of the body, spatial proximity, order, and visibility are required.

Not only the ordering of the body but also the ordering of subjective space is at stake here. The colonists aim to keep the Indigenous subject from wandering away physically, but they also aim to keep their subjective space within the domain of the Christian pastorate of care. The church should not be far from their bodies, and as a result it will not stray too far from their minds.

1512: The Laws of Burgos

The attempt to regularize and concentrate the habitus of the Indigenous subject in America is already clear, for example, in the first codification for the governance of subjects under the reign of the Spanish in America: the 1512 Laws of Burgos. The central concern of this 1512 document is the "idleness and vice" of Indigenous peoples and the difficulty of inculcating a productive, virtuous, and religious subjectivity when they live at such a distance from Spanish settlements. The subject may very well appear to accept the doctrines of the Christian faith and even begin to practice in a pious manner when in the presence of the Spanish colonists, yet as soon as they wander back home these doctrines slide right out of their subjective world. The Spanish are confounded by an apparent forgetfulness. The Laws of Burgos explain, "Although at the time the Indians go to serve them they are indoctrinated in and taught the things of our Faith, after serving they return to their dwellings where, because of *the distance* and their own evil inclinations, they immediately forget what they have been taught and go back to their customary idleness and vice."[25] The active force of forgetting is a continual threat to the production of an obedient subject.

The express concern is that the given spatial arrangement does not allow a completed subject formation: The distance allows the Amerindians to wander off back into their own subjective space of idleness and vice. Virtue and productivity can be habituated, however, if the Amerindian settlements are destroyed and relocated to the adjacent regions of the Spanish settlements. All this is justified, not in the name of the great power of the king or in some show of the force of Spanish colonizers, framed as a benevolent practice that will also reduce the hardships and health problems of the Amerindians. With the construction of a space that is said to produce a healthier, more benevolent condition, along with a virtuous and productive subject, we surely have the prototype of a disciplinary power formation.

Here we might recall María Lugones's argument that extends and critiques Quijano's argument about race and coloniality to include the construction of modern heterosexist gendered relations. Thinking together with both Lugones and Quijano we can see that the birth of a new anatomo-politics in the colonies is concerned with the production of both racial and gendered formations of the body.[26] In considering the disciplinary formation in the colonies, the extraction of forces from the body for labor must also be considered alongside the gendered relations of organizing and reproducing the family. The construction of the public-facing laboring body is also linked to the construction of private life of the family. The interdiction on public nudity for men and women could be seen as one such example of this public-private intersection.

The full-scale microphysics of anatomical detail described in *Discipline and Punish* are not fully developed at this point. Yet we do find the emphasis on the production of a subject as opposed to the restrictive or deductive model of sovereignty. The productivity of power is a key movement away from the violent excess of the sovereign's deductive use of force. The aim of bringing the Amerindians into immediate proximity of the Spanish gaze is to form a virtuous and productive subject who will yield more spiritual and physical value for the Spanish crown. Distance is a key problem that does not allow for sufficient supervision and shaping of the Indigenous subject. The space of wandering and idleness must be extinguished, and the Spaniards go so far as to justify the burning of their previous villages in the name of this subjective transformation: violence as a precondition of care. Although this question is principally posed in religious terms of conversion and virtue, there is certainly a parallel interest in the economic terms of productivity (increase of force) and political terms of obedience or docility (decrease in forces of resistance).

Amerindian idleness and wandering are opposed to the stable and rooted nature that the Spaniards wish to impress upon them: "The principal aim" of these practices on the part of the Spanish, as the document states, is that the "faith shall be *planted and deeply rooted*."[27] The desire for a stable and deeply rooted subject was a consistent concern across various Spanish and Portuguese colonies in the Americas. For example, Eduardo Viveiros de Castro writes of the sixteenth-century Jesuits in Brazil who were struck by the problem of the *inconstancy* of the Amerindian soul.[28] The problem for the Jesuits was not so much an active resistance to the teaching of Christian doctrine but instead a practice of forgetting carried out by the Amerindians. The Amerindians would readily accept the doctrines of the Christian faith and participate in its rituals only to turn around and forget them. As the Jesuit Antônio Viera recounts, "There are other nations, however—and such are those of Brazil—that receive everything that is taught them with great docility and ease, without arguing, without objecting, without doubting, without resisting. But they are statues of myrtle that, if the gardener lifts his hands and his scissors, will soon lose their new form, and return to the old natural brutishness, becoming a thicket as they were before."[29]

Thus, the Jesuits were struck with this problem that they referred to as inconstancy, the instability and mobility of the Amerindian soul that refused to have a stable relationship to Christian religious belief.

There is an overlap between the Jesuit theme of inconstancy and the problematic raised in the Laws of Burgos: how to produce a stable and rooted subject out of an inconstant and flexible one. These laws offer one of the first formulations of an emerging Spanish colonial urban order, but additionally they act as a kind of handbook for the daily, weekly, monthly, and yearly activities that the Amerindians will perform in their proximity to the Spanish. Every two weeks they will be examined "to see what each one knows . . . and to teach them what they do not know," also ensuring against the stubborn practice of forgetting.[30] After five months of working in a gold mine, Amerindians should be allowed to rest for forty days.[31] They are to confess once a year and whenever they fall ill.[32] Churches are to be built in all their towns and next to the larger mines, and if a church is too far from any estate, a new one is to be built. They are to attend Mass on Sundays and all holy days and are to eat their best meal of the week on this day.[33] The racial politics of consolidating Indian space also intersects with gendered practices of cultivating monogamous families and hiding one's nudity: The men are not to take more than one wife and must be repeatedly reminded of the evils of doing otherwise, and they are to be given one peso every year to buy clothing so that they go about sufficiently clothed. The public presentation of the disciplined body must also yield to this regime of gender and sexuality.

The pairing of violence and care also is clear if we reflect on how these laws were intended as a code for the good treatment and care of Indigenous peoples. These principles were established to reform and mitigate the worst effects of the economic system of the encomienda, yet they still served to uphold this system and uproot the Indigenous way of life.[34] In short, the point we can isolate in this discourse (which has been hailed by some as an early document of human rights) is the problematization of certain protodisciplinary techniques for the subject formation of a wandering or inconstant subject. These techniques involve the construction of a spatial and temporal world to govern, regularize, and examine the conduct of these subjects. These techniques also involve a racialization of space in that the Indian town is meant to be separate from, yet proximate to, the Spanish town—and the Indian town will gather people from diverse Indigenous cultures and languages and construct their habitus within a homogeneous grid of power.

This disciplinary space operates as a heterotopia, with the Americas as laboratory. This space acts as a counter space to those existing spaces in the center of European society. Europe constructs a blank slate to order new apparatuses of power and knowledge. The epistemic and political practice of colonialism is not simply one of exclusion, exteriorization, and ontological nihilism of the periphery; it is also a productive practice of power, an organization and control of subjects and space. Heterotopias inform the practice of producing new technologies of power and knowledge within new and different spaces. The emergence of the grid in this sense involves a heterotopic practice. It emerges out of an encounter with an *other space* and the practices built on the construction of this space as empty, organizable, and gridded. This empty space that can be ordered couples with a racial notion of an empty subject that can be produced: The Indian who did not preexist European colonialism in any sense (as category, race, or group) is produced in the disciplinary town.[35]

1573: Royal Ordinances on City Planning

The Laws of Burgos exemplify an early expression of Spanish colonial transformations of space and organization of subjects. These heterotopias engender the construction of protodisciplinary spaces outside of and different from those spaces of European societies. Yet these 1512 laws sketch only the very beginnings of a spatial transformation that would vastly increase and intensify as the Spanish reach expands across the Caribbean, South America, and Mexico during the sixteenth century and beyond. The disciplinary heterotopias of colonial cities become a central concern for the power apparatus of the Spanish crown and its viceroyalties, and their laws of construction more refined. These practices magnify the anatomo-politics of making bodies known, visible, and productive through the technique of the grid, which offers an ordered, knowable, and manageable citizenry.

The construction of the Spanish colonial city as a grid had many precedents in the sixteenth century, as we can glimpse from the Laws of Burgos along with King Ferdinand's 1513 letter to Pedrarías Dávila. By the second half of the sixteenth century, most Spanish colonial cities were built according to this model. Up until 1513, Spanish colonial settlement was primarily defined by failure and uncertainty with almost no real success in planting firm roots in their colonial cities. Yet, as these urban principles proliferate, especially following the 1521 conquest of Mexico, this would drastically change. In 1573, then, the rules for constructing a colonial city were given official codification with Philip II's royal ordinances on city planning.[36]

In these ordinances, the requirement of a grid design is formalized in detail along with the importance of placing a plaza at the heart of any city. Ordinance 110 of 148 states, "On arriving at the place where the new settlement is to be founded a plan for the site is to be made, dividing it into squares, streets, and building lots, using cord and ruler, beginning with the main square from which streets are to run to the gates and principal roads and leaving sufficient open space so that even if the town grows, it can always spread in the same manner."[37] The center or starting point of these gridded towns was always the plaza, which should be a square or a rectangle (ord. 112). In towns on the sea, the plaza would be at the start of the town but when inland, it would be in the center. This plaza was, furthermore, designed to be proportioned to the size of the town and future anticipated population. The plaza was not built to house any private residents but to be the commercial, legal, and religious center of the city: always complete with a church, government buildings, a bank, and space for merchants (ords. 119, 121, 126).[38] A hospital for the poor and for those with noncontagious diseases would be built close to the church in the plaza center, and one for those with contagious diseases would be built at the outskirts of the town (ord. 121). Businesses that produce considerable waste or filth such as slaughterhouses, fisheries, and tanneries are also to be positioned away from the center of town but in a space where the waste can be easily disposed of, for instance, alongside the river or sea (ord. 122).[39]

The town should also be designed in such a way that its grid can easily be expanded in the event of future population growth, and the buildings should be uniform in their design and appearance as far as possible, for the sake of beauty (ord. 134). No Indigenous subjects are allowed to enter the town until its basic construction has been completed.[40] This ordinance is designed to prevent any conflict while the city is being set up, so that once it is completed the Amerindians will recognize the firm roots of the colonists planting themselves in the New World and be less inclined to attempt to expel them or rebel. Here, the Spanish especially demonstrate the importance of their spatial technique of producing order: If they are able to successfully root themselves and the spatiality of their grid, it will be too difficult to uproot them and send them back to the shores from whence they came. Here, the principle of order and the grid is applied to the specific Spanish (and eventually criollo) town in a parallel way to its application to the Indian congregation. While the latter aimed to create a stable and disciplined Indian subject through this regime, the former aims to demonstrate the firm roots of European (and European-descended criollo) colonists: Both aim to secure the space against inconstancy and uprootedness through a stable order. Thus, the racialization of white European settler space in the colonial context occurs alongside the racialization of Indian space: They both follow principles of establishing a new order, linked and in proximity to one another.

Spanish colonization and conquest explicitly required a spatial technique of urban ordering to secure their claims and positions in the Americas. These ordinances laid out the methods for the Spanish to construct American space anew. Disciplinary spaces construct space and they operate according to their own rules and the distribution of roles within that space:[41] the placement of the churches, government buildings, hospitals, and commerce. In the colonies, space became a new problem, subject to new rules of construction: As a heterotopia, as a space freed from spatial entanglements of Europe, the Spanish constructed towns that could order a multiplicity of subjects according to a newly established logic. The construction of spaces anew also involves the construction of new modes of subjectivity and the racialized production of this subjectivity between colonist and colonizer.

Foucault points out how the geometric model of the town differs from the sovereign model, and he sees the Roman military camp as a prime influence: "In the case of towns constructed in the form of the [Roman military] camp, we can say that the town is not thought of on the basis of the larger territory, but on the basis of a smaller, geometrical figure, which is a kind of architectural module, namely the square or rectangle, which is in turn subdivided into other squares or rectangles."[42] Sovereign spatiality is based on the territory, capitalizing control from the center, and policing and preserving the borders at the limit of the territory. Disciplinary space is not tied to the logic and order of the territory but is produced on the basis of its own model or geometric figure. Indeed, among various influences, many scholars emphasize the Roman and Vitruvian influences of the grid structure of Spanish colonial urban design.[43] The Roman imperial project offered resources and inspirations to the Spanish architects managing an overseas empire.

Like the Roman project, the exercise of early modern Spanish power does not fit strictly within the nation-state model of sovereignty. It could also be said that sovereignty is itself a nascent and not yet fully formed principle in the sixteenth century. At the same time, however, Spain is one of the first emerging models for the possibility of an absolutist and unified sovereign territorial state in Europe.[44] In this light it is perhaps surprising to think of its colonial machine as one that produces disciplinary spaces. Could it be that the problematization of sovereign territorial space and constructed disciplinary spaces takes place simultaneously? The example of Madrid evidences this process. In 1561, twelve years prior to the ordinances on city planning, Philip II decided to move the capital city of Spain to Madrid. This decision was made according to a sovereign logic of territory: the need to place the government in the center of the state so that the sovereign's reach can extend across the territory and so that their power is felt radiating across it. As Foucault explains, "Sovereignty capitalizes a territory, raising the major problem of the seat of government."[45] This was the logic that led to the choice of Madrid as capital city. Madrid was an old medieval city, with winding roads and lack of any gridiron structure. However, Philip II would decide to re-design the center Plaza Mayor of Madrid to conform to his love for the rational gridiron structure. This would not be an easy process given the historical density of the organic development of Madrid's urban center. As Jesús Escobar points out, "The imposition of a grid atop the historic core of Madrid was not a possibility, and yet some of the theoretical concepts behind grid planning were actually carried out."[46] The desire for the grid to be placed atop medieval European towns points to the boomerang effect of bringing colonial styles of urban planning back to the metropole.

There are, thus, overlaps between the grid-structure planning of colonial cities and the desire for gridded plazas in the Spanish metropole. Yet the choice of Madrid as capital city and the redesign of its central plaza are not strictly disciplinary. The concern with Madrid was, first of all, territory. Second, the architecture of Madrid was intended to exhibit the strength and glory of the sovereign: to display a spectacle for all subjects to see. But the desire to impose the grid atop this sovereign space points to the clash and the complex engagement between sovereign and disciplinary power. Indeed, as Foucault suggests in "Of Other Spaces," the heterotopias of the colonies can be considered heterotopias of compensation.[47] The colonies compensated for the messy space of Europe, offering a new regime of order that was then imported back to the metropole and super-imposed atop the logic of sovereignty.

Furthermore, as discussed earlier, the construction of space in the New World was not primarily concerned with traditional military conceptions of defense in terms of the heaviness of walls and fortification, as is the case in sovereign conceptions of territory. Defensive concerns with respect to attacks from Amerindians were formulated instead in terms of gaining recognition of the colonists' stable foundations in these new spaces, showing strongly planted roots.[48] In short, the defensive technique of the wall was actually replaced by a productive technique of creating a predictable and virtuous subject alongside an organized and well-rooted space.[49]

The Persistence of Coloniality as Grid:
Race and Moral Geography

The *longue durée* of the colonial grid extends into the independence period in Latin America, where we see its persistence and failures. The nineteenth-century problem of ordering a newly born nation-state will be posed in terms of how to govern an expansive territory, including the unending countryside, on the model of a city. The early project of Spanish colonization involved less the attempt to produce ordered space across the entirety of the continent than that to create intensive cities of order and create a network of connections between them. There was a multiplication of the colonial city in various sites, with its grid and the attempt to bring the Amerindian from the countryside into the ordered space of the town, especially as noted in the Laws of Burgos.

As we saw earlier, the Amerindian congregation racializes space (and spatializes race) through grouping Indigenous peoples of diverse ethnic groups, customs, and languages into one homogenizing space. Within this space, disciplinary power takes effect as it works to instill new modes of conduct and subjectivity, to effectively produce the Indian subject. Both the Spanish colonial town and the Indian congregation are premised on the notion of producing order in the colony, yet there is a segregation of space between the Indian and the European (or criollo) that is meant to establish the hegemony of Spanish rule and ultimately the hegemony of whiteness in the colonies.

What happens when this project of order is extended to the entire inland territories of the Americas and then to the territory of the nation-state? The regime of order in the early colonial period is intensified only around a few crucial nodes, in a time when cartographic knowledge had mostly been limited to the coasts and ports. Inland exploration takes on a new impetus at the end of the eighteenth century in the late colonial period, prior to independence. Santiago Castro-Gómez describes this shift and expansion from earlier ordering projects:

> One reason that explains the sixteenth- and seventeenth-century hegemony of the Spanish empire was its ability to "striate the sea," i.e. to convert the Atlantic circuit into a "territory" where the circulation of commodities, slaves, and people between the new and old worlds could be perfectly regulated. But faced with the emergence of England, France, and Holland as new powers competing for control of maritime space, Spain was compelled not only to striate the sea but also the land, subjecting the physical space of the colonies to a strict organization of all flows. . . . The striation of space also, and primarily, included the extensive "mapping" of the population.[50]

Spain striates the sea with maritime power and, as I have shown here, also the colonial town and Indian village in this early period. The key shift Castro-Gómez marks is the expansion of this striating practice to spread across the entire land in the late colonial period and onward. If mapping landed space in the American continent takes on a new impetus in this time, so too does the mapping of populations. The science of space, geography, will unravel a new technique of constructing the anthropological place of the human and racially mapping the value and tendencies of different populations across the Americas.

The production of order in American space, thus, intersects with mapping populations: fixing different ethnic groups to moral and physical characteristics through a racialization of their geographic place. As Castro-Gómez shows, these racial concerns vary between geographic determinism and modified accounts of how human populations might adapt or acclimate to different environments as a plant brought from Europe to America. The geography of race is treated in this sense in parallel to questions of botany and agriculture.

These concerns of racial and moral geography do not end but only intensify in new ways with the birth of independent nation-states. No longer just the colonial outpost, the ordering of space extends to new problematics, the frontier of space to be ordered and civilized across the countryside. The countryside, in contrast, embodies fears of disorder, lack of civilization, and racial mixing. Order allows categorization and segregation spatially, so it acted as a backbone of racial consolidation in the colonial period. Order also produces disciplinary subjects and instills a specific behavior on a particular population: The fear of the expanse, however, is that populations are mixing, their behaviors uncontrolled, and cultivated by a rugged nature that leads to moral and racial degeneracy.

These concerns are crystallized in Domingo F. Sarmiento's seminal 1845 text on the physical and moral geography of Argentina, *Facundo: Or, Civilization and Barbarism*. According to Sarmiento, the natural landscape of Argentina, with its infinitely expanding pampa and the rural gauchos who inhabit it, poses a nearly unsurpassable obstacle to the march of civilization, which has taken hold only in the city: "Immensity is in all parts of the country: immense are the plains, immense are the woods, immense are the rivers, the horizon always uncertain, always blending together with the land, lost in haze and delicate vapors which *prohibit the marking of the point in the distant perspective, where the land ends and the sky begins*."[51] The geographic landscape and immensity of the Argentine countryside impede the ordering project from imposing any grid. One cannot clearly mark any perspective within this space: In fact, it is impossible to differentiate the sky from the land on the horizon. The ordering epistemology is rendered uncertain by this natural immensity. Parallel to the sixteenth-century problems of mapping and gridding urban space and the globe, the Argentine pampa is a smooth space par excellence: expansive, undifferentiated, and populated with intensive and unordered flows.[52]

For Sarmiento this physical geography, thus, also translates to a moral and racial geography.[53] The disorder of nature produces barbarism in the people who inhabit these spaces, and it also leads to race mixing through a failure to separate out different racial groups spatially. Civilization and order must be produced: They are not naturally provided by the landscape. Without proper order, the other is not yet produced as citizen, and the possibility of proper moral conduct is ruled out. More than that, the gaucho presents a threat as some combination of a mixed-race, poor, and undisciplined subject.

Sarmiento describes this problem as follows:

> The fusion of these three families [Spanish, African, and Indigenous] has resulted in a homogeneous whole distinguished by its love of idleness and incapacity for industry, as long as education and the exigencies of a social position do not spur it out of its customary pace. The incorporation of Indigenous peoples through colonization contributed a great degree to this unfortunate result. The American races [Indigenous peoples] live in idleness, and appear incapable of hard, unrelenting work, even when compelled. From this came the idea of introducing Black races into America, which has produced such fatal results. But the Spanish race has not shown itself any more capable of action, when abandoned to its own instincts in the deserts of America.[54]

The tendency toward idleness is attributed by Sarmiento to both race and space. The Spanish subject has become lazy through race mixing, internalizing these supposed anti-industrial traits of the Indian. Yet they also have not fared well in the pampa on their own: The lack of spatial order and the natural expanse of the geography also produce their own corruption and lack of civilization. There is thus a threat of a kind of geographic determinism of the Argentine pampa if it is not cultivated by the ordering project of civilization and, not only that, of a geography that can lead to racial degeneracy on its own with or without race mixing.

Sarmiento would later sharpen his critique of race mixing and squarely place blame on the lack of anti-miscegenation policies in Spanish colonialism for its lagging behind the white Americans in the United States to the north. Sarmiento's late work on race details explicitly racist views in a way that appeared more obliquely in *Facundo*.[55] Yet the degeneracy and the anxiety created by the pampa is one that was already clear in his earlier work: The backwardness of the gaucho is both racial and spatial, an unruly subject that cannot be contained by the disciplinary grid. In this we can see continuity with the early colonial project of ordering urban space and the attempt to discipline the Indigenous subject. The anxiety about the wandering Indigenous subject who shakes off their spiritual training when allowed to return to their own space is reminiscent of the anxiety of the gaucho abandoned to nature and disorder.

The disciplinary technique of producing a pious and industrious moral subject was concentrated early on within the city space, where the colonists could plant their roots and organize the daily life of Indigenous populations. In the postindependence nation-state, the question of how to impose discipline on an entire population, how to cultivate the proper moral conduct of an entire nation, does not translate to independence from the shackles of colonial techniques of power. Instead, we see the search for new avenues through which to transmit and generalize the disciplines and spatial ordering across the nation. Space becomes more abstract yet more precise in the eighteenth century. The desire to use cartography and to grasp the territory at a fine-tuned level is translated also into a racializing moral geography that maps the relationship between different places and the traits of different populations. Outside the productive grid lies the threat of a disordered nature, a source of degeneracy for the human subject. The Americas used to be outside and beyond the map of European knowledge, part of the uninhabitable antipodes of the earth that could produce only monsters.[56] Now, the American continent will be internally divided between spaces of order and disorder; the latter wait in potentia for their productive order.

2. Orders of Movement

THE TRAVELER AND THE SETTLER

Modern/colonial constructions of territory are defined not only by boundaries but also by movement and passage. Who can enter, who can leave, and who can move about? What kind of movement is permitted? In early modern European political philosophy this question is formulated, on the one hand, through the pairing of freedom with movement and, on the other, by asking what kind of relationship to land is necessary to constitute a property relation.[1]

Ordering space, then, is not only about fixed physical boundaries and sight lines but also the codification of space in terms of who moves where, when, and in what ways. The codification of geography, movement, and the subject forms a modern/colonial assemblage of space. The ordering and gridding of urban space not only sets up rigid boundaries but also enables the ordering of movement. To explore the intersection between dividing lines and porous movement, I take up two seemingly opposed spatial figures: the traveler and the settler. Both figures are produced and embedded within a moral geography of race that shapes the terms of who can move, who can settle, and who can forbid entry to others. In this sense, they are complementary rather than opposed figures—at times, two sides of the same coin.

For Indigenous peoples of the Americas since the conquest, land is emptied of rightful claims of habitat, and modes of relation, by strategies of colonization. Colonizers claim that the land is barren of legitimate modes of territorialization. They empty the land, claiming the Indigenous inhabitants lack adequate dominium over their own lands. Yet this racializing geography is not simply emptying: It operates in the shift between nullification and production. Claims of emptiness are conditions of possibility for producing certain kinds of subjects, movement, and order. Thus, the claim of emptiness also leads to new configurations of subjects, space, and conduct, as we see also in the case of shaping disciplinary space for Indigenous subjects.

Different subjects are racialized in different ways in this moral geography that starts by emptying the Americas. I focus particularly on the modes of dispossession of Indigenous space and the different modes of claiming possession of American space by Spanish and English colonists. By focusing on the contrasting dimensions of movement and settlement, I aim to show that there is not one simple mode of dispossession or racial-moral geography but various modes of territorialization that disqualify Indigenous and non-European modes of spatializing subjectivity. Thus, multiple techniques belong to this racial-moralizing of space, forming part of a larger technology of order (disorder) and dispossession. Racialized space thus sits at the intersections of a nascent ordered state form alongside a nascent mobile porous capital form.[2]

From the right to travel employed by the sixteenth-century Spanish theologian Francisco de Vitoria to John Locke's usage of the Americas in his theory of property and wasteland, we see that the settlement-movement dyad is articulated together with a moral-racial geography. As Jodi Byrd argues, we also see how the figure of the Indigenous is racialized through not only emptying logics but also logics of transit and movement, as they are continually displaced and pushed to an elsewhere at the edge of the map.[3]

In addition to the massive macro-level ordering of space, colonial legacies shape everyday modes of embodiment and possibilities for living space today. Here the question of resisting racializing geographies opens an approach to coloniality as persistent yet not all-encompassing. As María Lugones emphasizes, domination put into place over human subjectivity by spatial networks of power is always incomplete, and trespassing against the map of power is always a possibility.[4] Trespassing may refer to other modes of distributing oneself in space through spatial modes of resistance, going against the grain, or an art of refusal of the dominant logics of the coloniality of space.

Reason, Territory, and (In)habitability:
Who Can Move, Who Can Possess

Sixteenth-century colonialization strategically claims that a population's purported lack of reason negates rights to land, territory, and self-governance. As a precursor to more systemic racial geographies of later centuries, a geographically delineated group lacks reason and thus lacks rights.[5] The argument about reason and rights is deployed to define entire populations in new ways beginning in the sixteenth century. In this epoch, Aristotle's arguments on natural slavery from *Politics* take on an unprecedented importance, deployed to define all Native populations of the Americas. As early as 1510, the Scottish Dominican John Mair proposed that all Native peoples of the Americas are natural slaves in Aristotle's sense: lacking reason, proper forms of government, and written language.[6] With this global geopolitical application of Aristotle, he argues that the Spanish had a right and a duty to rule over the Amerindians and their territories. The claim that an entire population lacks reason thus leads to the essential forfeiture of all rights in the name of a natural hierarchy that is supposed to be mutually beneficial for colonized and colonizer.

Many thinkers developed and deployed Mair's usage of Aristotle's theory of natural slavery in the sixteenth century. The most well-known (and infamous) case is that of Juan Ginés de Sepúlveda in his debate with Bartolomé de Las Casas at Valladolid in 1550–1551.[7] Sepúlveda's arguments develop a moral geography of race in relation to the colonial project, as the entire continent of the Americas is regarded as a place that has not produced the proper moral and rational disposition in its people to count as full rights-bearing humans. Thus, Sepúlveda offers a model of hierarchical exclusion that racializes the population of an entire continent simply given its geography outside the Christian world (*res publica Christiana*) and its supposed degenerate conduct. On the latter point, Sepúlveda, who never traveled to America, is especially fixated on practices of cannibalism and human sacrifice.

From Mair to Sepúlveda, we see that early formations of race and racialization were much less about the physical characteristics of any human group and more concerned with questions about reason, moral and political conduct, and the modes of embodying these traits within a geography: Prior to the materialization of race in physical attributes, there is the spatialization of race and the racialization of space.[8] The Americas are outside any historical sphere of Western Christianity, and thus the people there are seen to be qualitatively different from those other infidels who have resisted or fought against the Western Christian religion and cultural norms. They are situated not as the infidel enemy but rather as the degenerate or less-than-human, a categorization whose implications are far more dangerous than that of an enemy worthy of battle.[9] In this sense, the religious space of the outside is quickly racialized into the degenerate space of the less-than-human. As Sylvia Wynter argues, race was the "construct that would enable the now globally expanding West to replace the earlier mortal/immortal, natural/supernatural, human/the ancestors, the gods/God distinction as the one on whose basis all human groups had millennially 'grounded' their descriptive statement/prescriptive statements for what it is to be human, and to reground its secularizing own on a newly projected human/subhuman distinction instead." As she notes, this newly projected human/subhuman distinction was predicated on the "Spanish state's theoretical construct of 'by-nature difference' between Spaniards and the indigenous peoples of the Americas."[10]

Movement and the Porosity of Territory in Vitoria

The argument of natural slavery applied to the Americas establishes a global abyssal space racializing the entire population beyond a certain meridian line as less than fully human. Beyond the line there is no reason, no legitimate mode of spatialization to lay claim to a territory: Thus, these spaces are emptied of the peoples who inhabit them and their modes of organizing their lives in space. We can see these arguments laid out in the 1493 papal bull *Inter Caetera* when Pope Alexander VI "donated" all the land west of a meridian line just past the Canary Islands to Spain for appropriation and all subjects therein for conversion.[11] Thus, what Boaventura de Sousa Santos calls "abyssal spatial reason" shapes much of what is to follow in Spanish and European colonialism.[12] Abyssal reason empties everything and everyone beyond the line not only of rights but also of humanity, reason, culture, and meaning.[13]

In this case, the abyssal logic is also religious and specifically applies to all non-Christian subjects. The title of "discovery" and the very notion of the "discovery" of the Americas would thus apply to this notion of land appropriation of non-Christian lands.[14] As Katherine McKittrick writes, "[These] spatial concerns of Mani became wrapped up in an ideological perspective that dehumanized and disembodied subaltern populations by conflating their beingness with *terra nullius*, places and bodies outside God's grace: idolaters in the uninhabitable; uninhabitable idolaters."[15]

However, the colonial project was not predicated on a purely abyssal empty-ing logic. The spatiality of abyssal lines in the papal bull of donation was not a sufficient justification. The spatial claim over the New World falters and a new principle of justification is sought. New modes of thinking this relationship be-tween reason and territoriality beyond the abyssal lines of discovery are called for, still in view of offering justifications for colonial rule and dispossession.

This history of colonial justification cannot be told without considering its great critic Bartolomé de Las Casas, who is supposed to offer an affirmative image of the Amerindian against Sepúlveda's hierarchical model of racial rule. Indeed, the great "defender" offers some of the most forceful condemnations of Spanish colonial violence against Indigenous peoples and is known as a great advocate for their rights. Yet Las Casas still racializes the space of the New World in a more nuanced manner. Against Sepúlveda's hierarchical exclusionary model, Las Casas's sees that "all mankind is one."[16] If Indigenous peoples have been only accidentally left outside the great flock of Christian subjects, the time has now come to bring them into the fold and expand the global reach of the pastorate. There is no intrinsic flaw to Amerindian rationality or spirituality, and Las Casas argues for the value of their institutions and practices. Yet he situates them as *potential* subjects ready to receive the word of the Christian God: meek lambs in need of salvation by entering the pastorate. Conversion through discipline and not compulsion.[17]

Las Casas humanizes the compulsion for the Amerindian to enter the Christian fold. Vitoria, instead, humanizes the rights of the Spanish to enter the Indigenous space of the Americas. Vitoria is often read together with Las Casas as an early sixteenth-century critic of Spanish colonialism in its most un-bridled forms.[18] In his 1532 lectures "De Indis," or "On the American Indians," he disputes the divine right of the pope to grant absolute dominion over non-Christian territories, seeking a more legitimate title of justification in the law of nations (*ius gentium*). The pope does not have temporal power extended across the whole globe nor can he grant such power to the emperor: "The emperor is not the master of the whole world."[19] Instead, Vitoria offers several arguments on alternative titles drawn for *ius gentium* and just war theory. The most sig-nificant principle he invokes is the right to travel (*ius perigrinandi*), which he relates to a right to society and commerce (*ius commercium*). Like Las Casas, he denies the claims of Mair and Sepúlveda that Indigenous people are devoid of reason, culture, and government. Despite this apparent affirmation of Indige-nous rights, Vitoria's principles of *ius gentium* and just war will legislate norms of reason disqualifying Indigenous relations to their space and denying their rights to self-determination.[20]

Vitoria's dispossession of space operates through notions of movement, porosity, and commerce: the Americas destined not only for settlement but also travel and commerce. Vitoria argues that this is a dictate of natural reason, a law that derives from the reasons of all peoples or nations (*ius gentium*): the prototype of what we now call international law. He defines *ius gentium* as "what natural reason has established among all nations."[21] This law has a somewhat ambiguous status, intermediate between divine law that bestows natural reason to these nations and positive law as it is enacted by diverse nations. When the positive laws of Indigenous peoples do not match the claims of natural reason declared by the European, the natural reason from abroad is seen to take precedence over what has been established in these Indigenous nations. Furthermore, this law has a strange status as it is supposed to govern the relations between nations while imposing certain norms of behavior onto them. When these norms are violated, it establishes legitimate grounds for just war. In other words, any potential sovereignty of Indigenous nations will always be compromised by the need to live up to this norm and, at the same time, rendered porous by the right of reason.

Several spatial notions are at work in Vitoria's claim that the right to travel is a part of *ius gentium*. First, he appeals to the notion that the earth was originally given to all humanity in common without any division of private property. Second, he argues that property was not intended to separate and fragment peoples from one another: "It was never the intention of nations to prevent men's free mutual intercourse (*hominum invicem communicationem*) with one another by this division."[22] Property is permissible and even desirable, but its purpose is not to put up uncrossable dividing lines. This principle is meant to ensure that some of the original in-common nature of the earth will be maintained even after divisions of private property are put into place. With respect to travel and the Americas, this will mean that impassable sovereign borders are not to be put in place against any foreign travelers who wish to pass through the land.

Third, Vitoria argues for the right to travel as a general principle by appealing to an analogy of other common spaces that have remained free and open to travel. Here the paradigm example is the sea. Under this analogy, American land is modeled to be like the sea: porous and passable, free of enclosure and constraints to passage. One commentator on Vitoria refers to this as the liquefaction of land.[23] Land becomes like the open sea, free and open to all modes of travel and all travelers at once. At the same, as I highlight later, this is the period when the smooth and endless space of the sea is subjected to a regime of order, such that it can be known, controlled, and navigated.

Land is liquefied, rendered porous, while the liquid flows of the sea are ordered and organized: How can we understand this pairing of the ordering of the sea together with the liquefaction of land? When we speak of the liquefaction of land, the Americas are seen as a porous space: not just empty in the sense of the 1493 papal bull but now porous. However, at the same time, the Americas, the globe, and the sea itself are being subjected to a new regime of ordering, setting up lines and divisions. This is evidenced in the global legal lines that divide between the claims of Spain and Portugal, the increased precision in mapping the entirety of the globe on the grid, and in the construction of Spanish colonial cities according to an ordered gridiron model.

The problematic of global space, of how Europe is to relate to the space of the Americas, is receiving one of its first theoretical formulations in Vitoria's text. In this light, we might consider Carl Schmitt's claim that "the new global image, resulting from the circumnavigation of the earth and the great discovery of the 15th and 16th centuries, required a new spatial order."[24] In a similar vein, Peter Sloterdijk argues that the sixteenth century commences a new process of "terrestrial globalization," a set of techniques for mapping, navigating, knowing, and ordering the globe.[25] Two irresolvable processes commence with this new spatial order: on the one hand, the movement of increased global commerce, porous boundaries that extend across the globe and know no possibility of enclosure, and on the other, the establishment of borders, boundaries, enclosed territories, and what will eventually become sovereign nation-states. Early on, globalization is already about the government of mobility in relation to the segmentation and ordering of space. Sloterdijk describes this as a process of abstraction in terms of flows of money and lines of geometry, reducing "all local particularities to the common denominator: money and geometry."[26] This problem could be parsed in terms of major problems like the relationship between globalization and sovereignty, or capital and the state. Here I wish to emphasize that the coding and racialization of space are at the very core of these emerging technologies of modern political and economic forms. What Marx calls primitive accumulation is predicated on a spatial apparatus that articulates together ordered, mobile, settled, and appropriable space alongside the subjects who have the rights to said movement, settlement, and appropriation.

In this sense, these spatial technologies are not separate but rather parts of one apparatus: Gilles Deleuze and Felix Guattari develop the term "apparatus of capture" in a way that could refer both to the striation of smooth space and the reimparting of the smooth onto striated (or ordered) space.[27] In this case, the striation of smooth space would refer to the ordering of oceanic space such that it could be known and navigated. The ocean is the ultimate kind of smooth space in that it cannot be fixed or physically bounded, and its expanse seems infinite, not unlike the Argentine pampa described by Sarmiento. Yet modern techniques of cartography and political techniques of seeking control and security would lead to an enormous effort to striate the ocean in the early modern period.

To reimpart the smooth on the striated would, then, refer to the liquefaction of (organized) land or territory: where Indigenous modes of ordering and distributing space are overlooked and replaced with notions of porosity and free movement. In this sense, sixteenth-century globalization commences, on a grand scale, a new worldwide process of the striation of smooth space (geometry), which will not only give rise to the networks that allow for the movements of global capital but also intensify and order the powers of an emergent political rationality that eventually takes on the state form. At the same time, globalization requires the porosity and free movement of colonizing subjects and the resources extracted from the colonies (money, or the silver of the new world). Here, Vitoria is a pioneer in articulating these relations among space, race, capital, and the legitimating grounds of law.

The Nomos of Global/Colonial Space

The territoriality of the emergent modern European state is determined by its colonial extension outward into supposedly free space, according to Carl Schmitt. He refers to this as the modern "nomos of the earth," a global spatial order defined by the bounded territories of Europe in relation to the construction of free, empty, and porous space beyond. Nomos is the foundation of law and legitimacy for Schmitt, and on a larger scale it is what organizes the earth. It is the origin of law itself. Schmitt traces the etymology of the term *nomos* to find that it originally refers to pasture and spatial division where law (and custom) are then built on these spaces of distribution, boundary, and enclosure. Furthermore, he describes the mythological language that is sometimes used to describe the earth as the mother of law, and he even appropriates these mythological notions in his own development of nomos. He explains that the fertile earth contains within it an inner measure, demonstrated in the way that human toil is rewarded with its fruits. Through the clearing of soil, cultivation of crops, and demarcations of fields and forests, natural lines and divisions of the earth develop. Subsequently, the order of social life is set up on the earth through delineation of fences, walls, boundaries, and enclosures. As Schmitt summarizes, the earth and the law are bound together in three ways: "She contains law within herself, as a reward of labor; she manifests law upon herself, as fixed boundaries; and she sustains law above herself, as a public sign of order. Law is bound to the earth and related to the earth."[28]

In contrast, "the sea knows no such apparent unity of space and law, of order and orientation."[29] Accordingly, Schmitt sees that the earth itself contains this inner measure and that law itself naturally springs from the earth. Even the human artifice of cultivating crops, organizing pastures, building homes, and establishing walls and boundaries proceed from this natural inner measure of the earth. The sea, however, contains no such inner measure or natural binding relationship with law. It is fundamentally a free and open space, one that "has no *character*, in the original sense of the word, which comes from the Greek *charassein*, meaning to engrave, scratch, to imprint."[30]

If nomos is a natural ontological ordering of the earth, it is also historical. Every era of the earth has its own nomos. Each nomos is a historically determined order insofar as the imprints of human labor and land appropriation on the earth change in each epoch. In fact, Schmitt explains that nomos is fundamentally defined and ordered through processes of land appropriation. Land appropriation is fundamentally the root of all legal order and the fundamental way in which states and human communities are ordered and divided. Thus, each era of the earth has its own principle of spatial ordering, its own nomos, and the modern nomos is one that came into place in 1492 and began to fracture, he argues, only in the middle of the twentieth century, post–World War II.

The divisions of the earth and the liquefaction of land in Vitoria are inextricable to this modern spatial order. For Schmitt, modern legal concepts and European international law are based on the spatial division of the earth after 1492. The modern nomos was constituted when the earth was divided between sovereign European territorial states, those with the possibility of mutual recognition and friend-enemy relations, and non-European spaces that were beyond the line and open for European land appropriation and conquest. What Schmitt fails to see, despite his revealingly open Eurocentrism, is the production of order that takes place beyond the line, in the space of the Americas. The Americas are not just the empty space of the liquefaction of land but also the space of the invention of order.

Gilles Deleuze and Felix Guattari develop on this notion of nomos in critical dialogue with Schmitt's account. According to them, nomos is spatial distribution prior to the order of the state rather than its ordering foundation. They write, "The *nomos* came to designate the law, but that was originally because it was distribution, a mode of distribution . . . without division into shares, in a space without borders or enclosure."[31] They affirm that nomos is originally a spatial concept of distribution but depart from Schmitt by arguing that it is without division, enclosure, or borders. The nomos is, instead, opposed to the ordered space of the polis, it is "the backcountry, a mountainside, or the vague expanse around a city," but it is never the ordered space of the city and law.[32] Nomos is the space of nomadism rather than the striated order of law and the state. In this sense, we can also think of Indigenous modes of spatial distribution that exist outside of and prior to the Eurocentric state-ordering principles as working under this wider, less rigid, sense of nomos. Yet these Indigenous spatial distributions are seen as empty by the European gaze, and they are nullified by this ordering project that seeks to both "striate smooth space" and reimpart the smooth on the striated.

In this account of nomos as distribution, there is no natural legitimacy to the state form and its borders that springs from the earth (as in Schmitt); instead the striation of space is a practice of the state, which embodies a capture and ordering of the forces at work in any given space.[33] Nomos as the space of distribution exterior to ordering of the polis is connected to the notion of smooth space, which is to say, smooth space is not empty or disordered space but space that is not divided and controlled by an abstract rationalizing project. The ordering and division of space by the state imparts the striation of space. Smooth space, then, might be a space of resistance to the ordering project, a space that goes beyond and outside the imposition of colonial ordering. In this sense, we must be careful to avoid the conflation of smooth space with empty space, where empty space is already caught up in the geometric-coordinate thinking that defines the striated. Smooth space is not empty space: Instead, it is populated with distributions that are not legible to state-centered geometric gaze.

As Deleuze and Guattari describe it, striated space is defined and determined by its grid; each point on the grid is defined by its position therein. The space is closed off by these dimensions and determined intervals, charting in advance any possible meaning and movements. Instead, in smooth space the point is without top or bottom, front or back; it is not ordered by the enclosure of the grid. Smooth space is directional: It follows a trajectory that is not charted in advance. When people inhabit smooth space, they distribute themselves within it rather than enclosing or rooting themselves. Examples of striated versus smooth space include the game of chess versus go; the order of the city (and state spaces) versus the open sea, ice, the desert, or the steppe; and optic long-range vision versus haptic (tactile, intensive, close-range vision). In this sense, the Americas are smooth space for the colonial gaze when they exceed the apparatus of capture of the ordering project. And smooth spaces pose a threat to the grasp of the colonial state when they cannot be ordered. The pampas of Argentina are smooth space in Sarmiento's account, insofar as they exceed the civilizing order that he wishes to impose. They become fully empty space only once they are gridded, mapped, controlled, and known.

Colonial empire deploys techniques of striation: It must striate smooth space to control and capture the forces that fall outside it or those that have yet to be ordered by it. It must chart the forces that inhabit its boundaries so that it may subjugate them and subordinate them to their coordinates as if a point on a grid. Meanwhile, smooth space is populated and traversed by forces that are exterior to the state, as are many Indigenous tribes' relation to space prior to its incorporation into the disciplinary village or the escaped maroon communities that form their own modes of distribution outside the reach of colonial power.[34] The subjects that inhabit these forces can distribute themselves and follow directional trajectories rather than being known in advance and captured by the state.

Schmitt also marks the distinction between the ordered space of Europe and an unorganized, free exterior. According to his claims, the source of law and the legitimacy of the state are based on land appropriation and the division, ordering, and bounding of said land. Opposed to the ordered nature of land appropriation and division, the sea poses a model of undivided and unapportioned space, as it is not possible to establish roots on the sea or to establish physical boundaries. The sea is without character in the strict sense, without imprints or cuts. The sea is in this sense a smooth space, one that is not easily or automatically striated. Yet it was precisely this modern nomos that began to striate the sea by making it known, ordered, and navigable.

Schmitt explains that a new spatial order of the earth developed after 1492 with the emergence of global linear thinking: the dividing and ordering of the earth in terms of meridian lines. The world was divided by these *rayas* (Spanish for "lines") in which whatever fell beyond the line was not subject to any traditional code of law. As Schmitt quotes from Blaise Pascal, "A meridian decides the truth"; on the other side of the line the rules of old Europe simply ceased to apply. He writes, "A closer juridical consideration of amity lines in the 16th and 17th centuries reveals two types of 'open' spaces in which the activity of European nations proceeded unrestrained: first, an immeasurable space of free *land*—the New World, America, the land of freedom, i.e., land free for appropriation by Europeans—where the old law was not in force; and second, the free sea—the newly discovered oceans conceived by the French, Dutch, and English to be a realm of freedom."[35] The cohesion of the old world would thus be established through a mutual balance of the right to freely appropriate land and sea beyond the line, on the one hand, and an internal order of Europe, on the other. By recognizing the outside as free for appropriation, Schmitt argues, European states were able to recognize each other as equals, or *justus hostis* (just enemies), that could be confronted on even footing.

However, following Deleuze and Guattari we can see what Schmitt leaves out of this account, which is the production of order on top of this supposedly free space. We could say that the problem is not one of balance between free and ordered space, as it is for Schmitt, but one of capture. As I showed earlier, the Americas are themselves the source and subject of an intensive regime of spatial ordering. A primary aim of the state assemblage is the appropriation and striation of smooth spaces. As Deleuze and Guattari write, "One of the fundamental tasks of the State is to striate the space over which it reigns, or to utilize smooth spaces as a means of communication in the service of striated space. It is a vital concern of every State not only to vanquish nomadism but to control migrations and, more generally, to establish a zone of rights over an entire 'exterior,' over all of the flows traversing the ecumenon."[36] While "free" space beyond the line offered up a region where very different lawless practices could multiply, the colonies also give rise to an intensive ordering of space and coding of who can move through it, to "vanquish nomadism" and "control migrations." In this sense, the grid formation of colonial cities in the Americas is a precise example of this practice of striating the exterior. Order is reimparted onto this land that has been emptied and liquefied. This order is coupled with the codification of who can move and who can travel: porous orders of space and the ordering of movement in the colonial/global world.

Reversals, or a Right of Refusal?

Returning to Vitoria, the openness demanded by his account of free travel might seem a right to hospitality in the positive sense. Yet his notion cannot be separated from the colonial and racial codification of such a right. Consider further, "Amongst all nations it is considered inhuman to treat strangers and travellers badly without some special cause, humane and dutiful to behave hospitably to strangers."[37] Taken as an abstract principle, this precept of *ius gentium* might be a valuable law to balance out the dividing lines on the map of power—if we were to apply such a principle today with respect to rights for migrants and refugees (especially those exercising a right to move from formerly colonized and war-torn spaces to historically colonizing centers of power), this might be the case. Yet in the colonial context the directionality of this spatial movement is reversed. The colonizer is claiming a right of entry into the land of the colonized based on a supposed natural relationship between reason, movement, and the earth's surface.

Vitoria goes on to point out that "all things which are not prohibited or otherwise to the harm and detriment of others are lawful. Since these travels of the Spaniards are (as we may for the moment assume) neither harmful nor detrimental to the barbarians, they are lawful."[38] And if this right to trade and travel is refused by the "barbarians," then Spain has a right to just war. No law of natural reason gives Indigenous Americans a right to resist this travel; no conception of Indigenous sovereignty is possible here.

Vitoria states, however, that his principle applies only if the trade and travel is not harmful to the receiving nation. Yet he claims that it can be assumed for the moment that this trade and travel is not harmful: a conceptual distortion resignifying Spanish colonists as peaceful travelers. Here what is left outside the map of power is most significant: Of course, Spain is not in America simply to conduct free commerce and travel with willing participants; it is there to extract resources, appropriate and settle land, evangelize, and exploit Indigenous labor. At the time of Vitoria's writing in 1532, millions of Indigenous people in the Americas had already been massacred by the Spanish by conquest, war, disease, and dispossession.

Travel is a cunning pretext for a colonial project. The traveler is the settler or the colonist in disguise: the subject who seeks access to any land, who has a right to enter and exit as they please, without regard to the impact it might have on the receiving party. Again, there is no international law or human right to refusal or resistance of this global extension of the map of power. The striation of American space works only on behalf of the colonial power; the smooth space is captured by the Spanish imperial project. When it comes to a notion of sovereignty over Indigenous space forbidding outside flows or reversing the direction of travel and entering Europe for free commerce, this possible blockage or movement is inconceivable.

This was not lost on all Vitoria's contemporaries. One of his greatest students, Melchor Cano, writes a pithy critique of his teacher in 1546. Stopping short of imagining Indigenous travelers entering the shores of Europe, he states, "We would not be prepared to describe Alexander the Great as a 'traveller.'"[39] Cano imagines the reversal of the spatial playing field, which illustrates the power relations at stake. The colonist cannot be described as a traveler when their aims are clearly not simply to pass through the lands of other nations to engage in peaceful communication and commerce. The history of imperialism within Europe makes the point for Cano that of course his nation would claim a right to refusal or defense against such an imperial threat: Their space would not be so porous. Cano critically exposes the map of power, unmasking the abstract spatiality of Vitoria's traveler for the concrete colonial subject at play.

The traveler is thus like the abstract subject described by María Lugones, who is "constructed as having deserved [their] spot," the "user of people's lives and labor without being touched by exercising [themselves] as apart from them,"[40] while the reverse movement of the Indigenous subject traveling on Spanish soil is hardly within the scope of Vitoria's imagination.[41] The concrete spatiality of colonial conquest is abstracted onto this map of power, where people are free to come and go as they like in a way that supposedly causes no harm. Lugones contrasts this colonial model of "unrestrained mobility" with her account of spatial resistance via what she calls "world"-traveling. Instead of the free mobility that encroaches on the space of the other and imposes one's world, world-traveling movement is made by historically colonized, racialized, and gendered subjects between different worlds of sense and space.

In one mode of world-traveling the oppressed subject must inhabit the dominant world of space and sense; they must fit themselves into this world, conform to this map of power. However, this dominant map of power is not the only world, according to Lugones, and it does not exhaust the possibilities of the subject: The resistant subject travels to other worlds of space and sense, builds coalitions, and trespasses against this dominant spatial sense. The colonized subject travels in a sense different from the traveler, who never really sees anything new.

Perhaps they also exercise the right to refusal in a different sense too: not necessarily a call of purity but a call for another mode of relationality to the land, territory, self, and community. It is a refusal to be trapped in the dominant logics of the coloniality of space that offer no alternatives to Western regimes of the state and private property. Glen Coulthard articulates such a logic of Indigenous resistance as a refusal of state-based recognition projects.[42] The logic of identity/difference of recognition does not offer an alternative to the coloniality of space but only the possibility of inclusion within a project that erases an Indigenous relation to land and place. The inclusion-recognition dyad is governed according to the dominant nomos, not according to an alternative mode of distribution in relation to the land. In the Caribbean context of the coloniality of space, Édouard Glissant refers to this right of refusal to recognition as a right to opacity: a right not to be grasped and consumed by colonial logics of understanding.[43] It is a right to another world of intelligibility and another way of being. Ben Davis emphasizes the use of a rights-based language to make this claim: not at the juridical level but rather at the strategic level of struggles and resistance-based rights claims.[44] Perhaps we can think of a contemporary resistant and strategic reappropriation of Cano's critique of Vitoria in this light: from Indigenous resurgence to Caribbean creolization.

The Settler

If the traveler is thus a colonizer in disguise, it is not surprising that their movement is also inseparable from planting roots as the settler does. While the former operates on a principle of movement and porosity and the latter through rootedness and enclosure, these two figures work in concert rather than in opposition with each other in the modern/colonial paradigm. Movement across the abstract ordered map and the enclosed rootedness of a specific place are often coordinated together, striating smooth space and reimparting the smooth on the striated.

Colonization and conquest involve more than perpetual movement; the moving root must be transplanted to settle and spread in the New World. Glissant offers an image of the colonizer, in this sense, as the moving root, the process of transplantation to a new space.[45] The image of planting and roots is, of course, very prominent in various colonial discourses. It is clearly evidenced in the words of Puritan settlers in the late seventeenth century, as they suffered uprooting after the Indigenous uprising of Metacom's rebellion (King Philip's War). In the words of William Hubbard from 1676, "When God first brought this vine out of another land . . . he cast out the heathen, and planted it, he caused it to take deep root, and it was ready to fill the land; the hills began to be covered with the shadow of it, its boughs began to look like goodly cedars: it might have been said in some sense, that we sent out boughs to the Seas, and our branches to the rivers: But now we may take up Lamentations following, Why are our hedges broken down, and the wild boar out of the wood doth waste it, and the wild beast out of the field doth devour it?"[46] The image portrays what Glissant calls the colonizing rootstock: a singular ethno-cultural-linguistic root that spreads across a land and consumes all modes of difference and resistance that stand in its way. The Puritans envision their New World settlement as a mode of transplanting roots across space and time. Once successfully planted in the New World, this root is supposed to take hold, expand across the land, and eventually establish firm borders or hedges of protection. The expanding rootstock is thus often coupled with the image of the hedge that surrounds and protects the white settler from the outside chaos of the untamed wilderness.

Much earlier, after the American Indian epidemic of 1616–1617 killed around 90 percent of the Native population in Massachusetts Bay and forced most of the survivors to flee inland, there was a great deal of space "opened up" for the colonists.[47] The idea of a "wilderness wasteland" gave a conceptual determination to the land that had been "opened up" for the Puritans in America. The Puritans would consequently claim that this area was a wasteland because it lacked inhabitants, cultivation, and enclosure. In John Cotton's 1630 sermon "God's Promise to His Plantation," he speaks of three ways in which God makes room for a people. By victory in war, by charity (or purchase), and finally "when he makes a Country though not altogether void of Inhabitants yet void in that place where they reside."[48] Cotton takes solace in the double premise that it is not a land altogether void and vacant of inhabitants (which would perhaps hold malevolent portents of being uninhabitable) and that it also now has a great void of space that can be inhabited. That people previously distributed themselves within this space and lived off this land indicates habitability. Yet the land has been opened for the Puritan through the "fortune" that God has made space through an epidemic.

This vacancy alone, however, does not suffice to justify the occupation and ownership of this land in the Puritan mind.[49] The supplementary premise, which reemerges centrally for Locke, is that one can gain ownership over land that previously laid waste by cultivating it and causing it to bear fruit. Cotton says, "Nor doth the King reject his plea, with what had he to doe to digge wells in their soyle? But admitteth it as a *Principle in Nature*, That in a vacant soyle, hee that *taketh possession* of it, and *bestoweth culture* and *husbandry* upon it, his Right it is."[50] The production of emptiness, whether through epidemic, conquest, or conceptual violence, is coupled with an added requirement of producing order through culture or cultivation.

While the American wilderness is seen as a sanctuary from the imminent collapse of the religious order in England, it also poses a threat to their settlement. The Puritans see not only their land claims but also their religion as a protective hedge, drawing an enclosure around them to keep safe from harm.[51] The settler seeks to organize space such that they are secure from outside threats and sanctified within—they seek to construct a new order in the wilderness.[52]

Around 1672, prior to the shattering of their sanctuary by the 1676 uprising, the Puritans see the wilderness in the following series of registers: (1) a solitary place without inhabitants; (2) an uncultivated land without comforts; (3) an unpaved land without any paths, destined for wandering; (4) a dangerous land filled with threats of evil and temptation; and (5) an unenclosed and unhedged area without defense, subject to injury of whoever shall decide to pass through it.[53] The first three notions point to the settlers' desire for planting, cultivation, and marking out of paths and fences: what Locke will call enclosure. The latter two, however, highlight the increasing threats that Puritans felt from Indigenous resistance, and they mark them in response as almost inhuman for subsisting under these wilderness conditions—namely, wandering around uncultivated land without adding paths, enclosures, or husbandry.

Puritan moral geography racializes the Native relationship to space as evil and wayward. For the Puritans, upright morality was tied to the cultivation of land and the structuration of an enclosed space, the forming of a hedge. Because the Natives did not fulfill this requirement, their wilderness condition was seen to be without culture or industry.

Finally, because they considered the Native to be without culture, pathways, cultivation, or enclosure (and their difficulties in converting them to Christianity), it was surmised that they had made a pact with the devil.[54] The wilderness is no longer simply the unshaped earth but animated with evil forces: forces that are both absorbed by and infused from the Native inhabitants. The Natives are entrenched in evil because they had no enclosed land to cultivate their culture and their soul and were instead subject to wandering the uncharted wilderness, not unlike the images evoked of Native peoples by the Jesuits in Brazil. That is to say, the Natives are not considered evil in and of themselves, but extended time under this spatial regime of the wilderness required conversion to an ordered and enclosed space. The Puritans are unable to perceive another world of space that belongs to the Native; they are unable to perceive that they have their own modes of charting paths and moving through the space of this supposed wilderness.

Locke's *Second Treatise of Government* evidences a similar moral geography of American Indian land concerned less with evil forces haunting the wilderness and more with the problem of waste pitted against industrious subjectivity. Locke sees uncultivated land as wasteland and explicitly connects his abstract theory of property to the example of the hunter-gatherers of America. For Locke, the Native relationship to land is objectionable because it does not utilize enclosure and cultivation, at least not in ways that are recognizable to European modes of perception. Failure to cultivate quickly translates into waste for Locke, as he sees no Native nomos or distribution on the land and sees only that the land lies fallow. This supposedly fallow land does not bestow a plethora of fruits unto mankind. As Locke insists, "Land that is left wholly to nature, that hath no improvement of pasturage, tillage, or planting, is called, as indeed it is, *waste*."[55]

Locke's insistence that natural reason requires humans use land for the purpose of cultivation once again imposes, similarly to Vitoria, a norm of how human subjects should relate to and distribute themselves on a given space. For Vitoria the norm is one of movement and porosity, while for Locke it is enclosure and cultivation. Locke gives force to the notion that these are moral goods, that to leave land lying waste is a sin against humankind. In a cunning reversal of the problem with which he began his treatise—namely, that earth was given to all in common (the same problem that Vitoria cites at the outset of his account of the right to travel)—he conceives of the usurpation and cultivation of fallow land (private property) as a greater good to humanity: "He that incloses land, and has a greater plenty of the conveniences of life from ten acres, than have from a hundred left to nature, may truly be said to give ninety acres to mankind."[56] Not only does the Puritan do a favor to mankind by cultivating land, but the American Indian does a disservice by not increasing the fruits and luxuries that the earth might bear. Here, the racialization of geography is not about physical features of white or Indigenous bodies but instead about the relationship between the body, space, and land.

For the Puritans, the Indigenous way of life is corrupt because of their wilderness condition and their wandering without an enclosure. Locke goes further to think of a kind of necessary relationship, based on natural reason, between one's body (labor), land (cultivation), and the industrious subjectivity tied to such an ordered space. Once again, the Indigenous mode of spatialization is both erased and disqualified. In this case, it is the Native who shouldn't wander or journey, unlike the free-leisurely movement of the travel-colonist in Vitoria. These norms of settlement and travel are thus codified in relationship to specific populations and, in that sense, racial norms. In both cases, we see the movement and spatial distributions of the colonized circumscribed by a colonial map of power. Both demonstrate a moral geography of race, in which the moral status of the Indigenous is negated through their relationship to space, either in the form of welcoming the colonizer into the space or in the sense of a requirement for enclosing the space.

Conclusion: Trespassing and Resistance

Thus, from Vitoria to Locke, from the traveler to the settler, we have two distinct yet mirroring modes of racially coding spatiality, which establish the orders and roots of the colonial state form alongside its porous economic flows. Both cases foreclose and delegitimize other modes of subjectivity and space, apart from the spatiality of the colonizer. Vitoria's account lacks any notion of sovereignty granted to Indigenous nations, rendering them porous to the colonizer. And his claims will also be essential to establishing the porous flows that define the Middle Passage and the movement of Africans forcibly deported to the New World on the slave ship. With Locke and the Puritans, the moral geography is dismissive of a certain mode of space, rendering Indigenous hunter-gatherers morally deficient and incapable of claiming a right to the land they inhabit, as it goes wasted without cultivation and enclosure. There is no nomos aside from the nomos of the colonizing imperial state, and thus these two norms of space mark the coloniality of space.

The coloniality of space in these examples involves a flattening of space, an abstracting of one mode of spatiality that becomes the only mode of correlating reason, subjectivity, and land. This kind of spatiality is unitopic, only one world of space to be inhabited, a flat space that one must move within.[57] In tracing this imposition of unitopic space, I have sought to articulate the macro-level order of striation together with the micro ordering of subjectivity.

María Lugones describes this oppressive and fine-tuned dimension of the coloniality of space when she writes, "Your life is spatially mapped by power. Your spot lies at the intersection of all the spatial venues where you may, must, or cannot live or move. . . . There is no 'you' there except a person spatially and thus relationally conceived through your functionality in terms of power."[58] All the possibilities of the subject are laid out on this spatial map, where and how you can move, set out in advance. Lugones captures how this map is particularly coded and "highly restricted" for the lives of the oppressed and historically colonized, especially in terms of race and gender: Your place is fixed, she says. In contrast, the lives of the powerful move about on a freely charted course: "You may go places as a boss, pleasure seeker on the labor of others, tourist, colonizer, and user of people's lives and labor without being touched by exercising yourself as apart from them, and you are ideologically constructed as having deserved your spot."[59] By focusing on the figures of the traveler and the settler, we can see how this map was constructed and how moral-racial geography constructs space in such a way that some are placed in the position of having deserved their spot, while others are highly restricted in the course they can chart or taken off the map altogether.

Lugones goes on to point out, however, that this "you" who is charted by the map of power is only an abstract you, one that does not capture the embodied experience of spaces and the possibilities for transgression and resistance to the map of power. Thus, Lugones's account of space points us to the possibility of "trespassing against the spatiality of oppressions" and "redrawing the map."[60] The ways in which people are fragmented and charted by power is never complete. Coming to recognize the ways in which one is charted on the map is already to begin to open the field of resistance, where resistance can be very focused in small moments of subtle tensions brought to the fore. Even if the macro-level map is not redrawn, one can build coalitions between multiple worlds of nondominant sense through these resistances that exceed the map's capture. These gestures take a step beyond the abstract map through micro resistances and trespassing.

If we avoid thinking of space as flat or of colonization as an all-consuming project without remainder, we find that there are multiple worlds of sense and space that fall below the radar of the unitopic model. Lugones writes, "As colonized and racially gendered, we are also other than what the hegemony makes us be. . . . If we are exhausted, fully made through and by micro and macro mechanisms and circulations of power, 'liberation' loses much of its meaning or ceases to be an intersubjective affair."[61] Accordingly, there is not simply one hegemonic world that subsumes and dominates all subject positions, but there are many different worlds that one can inhabit, including the construction of a nondominant and resistant space-time. These multiple worlds point to different possibilities of embodying, living, and constructing space. As we see above, to Lugones this is not just individualized practice but an "intersubjective affair" that builds coalitions between different resistant worlds of sense and space.

Unlike the substantive "traveler" in Vitoria, the gerundive practice of world-traveling in Lugones's account always involves processes of openness to self-transformation as one connects to other worlds through loving perception. There is leeway to travel in this alternative nondominant sense.[62] The colonial traveler or tourist, however, does not really travel to a different world. They bring their world with them and seek to plant it anew. They are not transformed by this travel, nor do they seek to open onto a different world. Instead, the colonial traveler seeks to transplant a root from one place to another, enclosing and hedging in a world without seeing things otherwise. As Glissant says, the colonizer's "arrow-like nomadism" is only a disguise for "a devastating desire for settlement," not a true journey.[63]

While, in another sense, the colonizer *is* transformed by this travel and settlement. A new regime of space *is* produced in these new world conquests. Yet the colonizer does not recognize the specificity of their unitopic imposition; instead this spatiality masquerades under the banner of universality on the abstract map. These are the norms that all nations must agree to in terms of a natural relationship between body, movement, labor, land, and porous borders, ones that place the white colonizer in a privileged position while disqualifying Indigenous and African rights to space.

While Vitoria's traveler sought to impose their spatial regime on all others in a unitopic model that renders all space porous for their travel, Lugones's model of traveling between "worlds" does not require traversing the seas and the continents but instead finding and building spaces outside the dominant world of sense. All colonized peoples have been forced to travel to other worlds of dominant sense in which they are constructed as inferior and yet must make the effort to fit in within this spatial world, to construct a self that fits into this world. But this dominant world does not exhaust the world, it does not consume the subjectivity of the resistant subject. In her essay "*La Callejera*/The Streetwalker," Lugones describes the tactical spatial strategy of "the hangout," to build a spatiotemporal world beyond the abstract space that is imposed on the city. Creating a time and a space that allows for coalition building and allows for multiple worlds of nondominant sense to enter conversation. The hangout might also be a sense of trespassing against the unitopic model of space inherited from colonialism. As she writes, "Trespassing against the spatiality of oppressions is also a redrawing of the map, of the relationality of space."[64] In the hangout one finds interstices of space that are not captured by dominant sense, one connects with other worlds of nondominant sense in this "intersubjective affair." Ultimately, trespassing does not require an escape to some pure beyond colonial unitopic space. Instead, what is at stake is the carving out of the space of different worlds and the recognition that one is not exhausted by the dominant sense of space, left over from the colonial map. Another distribution of space, another nomos is possible here.

This mode of trespassing against unitopic regimes of space also invokes a kind of right to refusal and right to opacity with respect to the dominant juridical-colonial organization of space.[65] As Coulthard emphasizes, Indigenous recognition does not capture practices of Indigenous resistance as they seek to articulate another logic of relationality with the land. Just as the Puritans disqualified the Indigenous "wilderness" relation to land and place in the seventeenth century, contemporary dominant spatial organizations continue to overlook Indigenous, Black, Latinx, and historically colonized groups' claims to other modes of spatializing the world. As Audra Simpson points out, with respect to practices of Mohawk "nested" sovereignty and their refusal of US or Canadian citizenship: "There is a political alternative to 'recognition,' the much sought-after and presumed 'good' of multicultural politics. This alternative is 'refusal.'"[66] This refusal opens the space to another world and contests that notion that the matter is "settled" or finished;[67] other nations and other worlds still exist within and among the dominant colonial order.

PART II. Transmodern Cartographies

3. Transmodernity and the Battlefield of Coloniality

Go along the roads and through the fields and *compel*
them to enter [*compelle intrare*] and fill my house.
—JUAN GINÉS DE SEPÚLVEDA, ca. 1550

The caesura that establishes the distance between reason and non-reason is the origin.... We must speak of these repeated gestures in history, leaving in suspense anything that might take the appearance of an ending, or of rest in truth; and speak of the gesture of severance, the distance taken, the void installed between reason and that which it is not, without ever leaning on the plenitude of what reason pretends to be.
—MICHEL FOUCAULT, *History of Madness*

The *longue durée* of coloniality is founded on a spatial ordering of the globe. The project of putting Europe at the center, establishing the spatial conditions of Eurocentrism, also materially produces the global periphery. The spatial ordering and totalization of the globe provides a frame to consider other dimensions of the coloniality of space that I have discussed in terms of emptiness, movement, order, settlement, and discipline. Discussions of this global totalizing process have especially been at the core of Enrique Dussel's philosophy of liberation since at least the 1970s. The "geopolitics of knowledge" of European colonialism placed Europe at the center of the world and situated all other regions as the periphery tied to a lower rung, or entire lack, of knowledge and being.[1] Dussel's philosophy is a spatial philosophy, and his earlier work is especially influenced and organized around a spatial critique of knowledge and ontology using the terms of center and periphery. These global spatial concepts are inherited from earlier Latin American innovations in global political economy in dependency theory that explored the notion of systemic and structural underdevelopment in the global south in an explosive and transformational way.[2]

The spatiality of coloniality at the global level involves a double move of totalization and division, global enclosure and expansion, on the one hand, and global hierarchy and separation between who is considered human and nonhuman, on the other. A global network of circulation of commodities, culture, religion, and bodies forms for the first time, while this global network also stratifies spatial nodes of power between center and periphery. Coloniality involves a relation of nonrelation, a relation that totalizes while foreclosing and racializing the spaces of non-European peoples.

In this chapter's first epigraph, Juan Ginés de Sepúlveda gives us an image of this relation of nonrelation, the gesture of separation that is at once a compulsion to enter. It is strange, perhaps, for Sepúlveda to cite this passage from scripture in favor of evangelization of Amerindians when he is the same thinker who questions their very humanity, rationality, and capability for religion. Why would he want to bring subjects into the fold that he does not find to be capable of his standard of humanity? The key term is *compelle*, supposed to offer scriptural justification for the use of compulsion or violence to complete the task of evangelization. The emphasis is less conversion itself and more the justification of hierarchy and violent power over the Amerindian other. And at the same time, the movement of entrance is also a spatial gesture, the foreclosure of the outside, the expansion of totality.

In the 1512 Laws of Burgos, we get another image of this relation of nonrelation. The Indigenous subject slides out of the subjective world of the Spanish settler and forgets their indoctrination. There is a gap that these laws seek to close, a distance that needs to be shortened to bring the Amerindian into the flock. Thus, the laws propose that Indigenous settlements be built close to the newly established Spanish colonial towns. Once again, this is a sort of compulsion to enter. The distance between the Spanish colonial settlements and the Indigenous preconquest modes of living and distributing space allows too much space of difference; it allows the cultivation and maintenance of difference in the face of this totalization.

It might seem that the compulsion to enter runs against the caesura of separation described by Foucault in this chapter's epigraph from *The History of Madness*, where he traces another dividing line of modern European reason and its irrational other. How is the spatial proximity established by Spanish colonial settlers also a mode of separation and the establishment of nonrelation? In Enrique Dussel's terms we could describe this spatial closeness of the Laws of Burgos or the compulsion to enter as a form of *proxemia* instead of *proximity*. Dussel reserves the latter to refer to ethical relationships between human beings. *Proxemia* is instead the closeness that one has with objects, or in this case, the closeness when one is treating a human subject as an object.[3] *Proxemia* is the closeness of instrumental reason, of grasping and manipulating objects. This is the kind of closeness the Spanish seek to attain over the Amerindians, a closeness that will allow them to inculcate their mode of life and subjectivity: to monitor, discipline, and convert, or for Sepúlveda, simply to rule over. *Proximity* in Dussel's sense is instead an ethical mode of openness that one has toward others—it involves a preconceptual sensibility, the possibility of another mode of sensing and relating that is not foreclosed by the subject-object binary.[4] This other kind of spatiality of *proximity*, the possibility of another mode of sensing, thinking, and connecting with others, will be crucial to the notion of transmodernity.[5] Transmodernity is another kind of spatial opening beyond or across modernity, beyond separation and enclosure, toward relation and plurality: not the placeless universal but the situated pluriversal.

If modernity was constituted spatially as a sort of global battlefield, it is inextricable from coloniality. It was constituted through a process in which the periphery was silenced, and Europe became a unique site of universality. Enrique Dussel lays out this critique of modernity, resituating the modern paradigm globally to emphasize the colonization of the Americas as the central determination of this history. Here, I draw from and sometimes depart from his work in order to develop a notion of global spatiality as a battlefield, a materialist reading of the other, and the possibility of transmodern modes of thinking, sensing, and relating in the wake of coloniality.

Mythmaking Modernity and Inventing the Americas

The invention of the Americas materially produced the epistemic and ontological closure of non-European ways of knowing and being. And these processes of closure and exclusion were intrinsic to the establishment of European universalization: the process through which Europe takes its particular forms of knowledge, culture, and religion and claims them as universals for the whole globe to recognize and inculcate. Modernity, in this Eurocentric sense, is the process whereby the particular is claimed as universal, its spatial location forgotten as particular, and consequently spread out on a worldwide scale. To critique the mythology of this forgotten spatial conquest and expansion, following Dussel I situate modernity here as a global battlefield constituted in the planetary horizon of the world-system that was possible only after 1492. This latter sense of modernity accounts for its spatial constitution within a global field, as opposed to a unique and miraculous event that occurred internal to the geography of one particular place.

What Dussel refers to as the first stage of modernity[6] was one of geopolitical domination by Europe as the center of the first world-system alongside the epistemic and ontological universalization of European knowledge and existence.[7] In the second stage of modernity, this domination continues, but its constitution through a planetary process is forgotten. The process of forgetting enables Europe to provide a self-definition of modernity unrelated to its expansion on a worldwide scale. Europe understands its knowledge and culture as the unique universal but forgets to consider the history of geopolitical, epistemic, and ontological domination that was built into the construction of such a universal. This active process of forgetting is a key force in the construction of the myth of modernity, which involves the forgetting of the planetary happening of modernity. It is a myth that consequently relegates the outside and the periphery to nonbeing, sub-being, treating these peoples and places as ones that require development, exploitation, or destruction. It is the establishment of the caesura that renders the cry of the other inaudible to the dominator.

The mythmaking of modernity involves the material covering over and silencing of the other. It attempts to deny any epistemic or ontological position outside the one that has been universalized by Europe and, thus, establish the relation of nonrelation. When Europe forgets the spatial dimension of modernity as process, it forgets that universality, which is claimed here on a worldwide scale, is possible only because of the worldwide geopolitical extension that Europe developed during the long sixteenth century (1450–1650).[8] Instead, transmodernity opens up modernity to its planetary horizon and considers a plurality of possible epistemic sites of enunciation and styles of being.

The account of modernity provided here is intended to critique the blindness toward space in the standard accounts given by Max Weber, Charles Taylor, or Jürgen Habermas, among others. It aims also to extend and deepen the spatial account of modernity provided by post-structuralist critics of modernity like Michel Foucault or Gilles Deleuze. For example, in Foucault's treatment of the history of exclusions (madness, the great confinement, and the emergence of psychiatry) and enclosures (factories, prisons, schools, hospitals, and the emergence of modern discipline) and his spatial method of analyzing histories of domination, it is surprising that this analysis rarely extends beyond Europe. Foucault signals the importance of doing this work in places, as when he mentions in a footnote to *Discipline and Punish* that the colony or slavery could have just as easily been examples of disciplinary power. Similarly, Deleuze and Guattari look at relations of smooth and striated space along with the battles between nomads and the state, but these reflections could be further developed with an extended consideration of the striated ordering of space involved in the history of colonialism. Consider further the 1440 date associated with the "Smooth and the Striated" chapter in *A Thousand Plateaus*, which is very suggestive, as this date marks the commencement of the Portuguese slave trade and their exploration of the West coast of Africa. This date and its relation to the development of global coloniality goes undeveloped, however, and remains at the level of mere suggestion. These most radical critiques of modernity forget to consider the spatial forms of domination that emerge globally and their relation to colonialism. The result of this forgetting is that even the excesses and failures of modernity are situated internal to Europe without a sense of global entanglement.

Space as Battlefield and Transmodernity

To demonstrate an account of modernity that works with and through the spatial history of coloniality, I emphasize the spatial sensibility of Dussel's thought here. His work excavates the global battlefield of modernity alongside the myth of modernity that forgets this planetary horizon. Dussel's thought is situated in the global battlefield of modernity that continues to effectuate force on the creation of thought and struggles of resistance today. He writes his philosophy as a barbarian philosophy from the underside of modernity, and he considers his philosophical task to also involve a practical orientation of struggle.

Dussel's thought emerges within a field of struggle. He escaped to exile in Mexico City in 1975, leaving his home in Argentina where radical thinkers and activists faced deadly persecution in the buildup to the 1976 military coup. He writes *Philosophy of Liberation* (published in 1977) without the help of his personal library, which was partially destroyed in an attack on his home in Argentina. Instead, he crafts the concepts of this book from memory and without citation. He is uprooted as a thinker in exile, and yet he is engaged in philosophy precisely as a struggle against the attempted oblivion of radical liberatory thought.

We read something of this real field of struggle when he writes in the opening pages of his book, "Space as a battlefield, as a geography studied to destroy an enemy, as a territory with fixed frontiers, is very different from the abstract idealization of empty space of Newton's physics or the existential space of phenomenology."[9] For Dussel, it is clear that philosophy takes place in such a battlefield and that one's position within the battlefield cannot be dismissed as mere epiphenomenon. Concepts are carved up within these boundaries and their architecture is not innocent. They can be organized and mobilized as forces of domination or liberation. He even argues later in the chapter that when concepts tend to mobilize the forces of the center and protect one region at the exclusion of a periphery, they become forces of domination. Concepts that emerge from the periphery, however, have a critical perspective of the prevailing system and can sharpen this perspective as liberatory forces.

In modernity, the boundaries drawn have increasingly consolidated around the division between a center and periphery. The creativity and the very life and reason of the periphery was foreclosed, while the domination of the center became reified. Eurocentric thought shaped itself with the mission, not always conscious and often unconscious, of offering a justification of the center and its universalization while excluding the periphery. In fact, it was only in the first period of modernity (1492–1640) that there was an explicit consciousness of the question in which the justification of the center's power over the periphery was raised as a philosophical problem. All the major sixteenth-century Spanish thinkers seriously address the questions of the justice and injustice of Spanish colonization, as we see in the cases of Las Casas, Sepúlveda, and Vitoria.[10] Subsequently, the question of domination and conquest goes unquestioned, and Europe's position as center of the modern world-system shapes the horizon of what can be seen. The question to ask from the underside of modernity is how the epistemic frame of modernity has made one see in this way and how it might be possible to shift the horizon of this vision.

Philosophy and one's position in a field of conflict are always bound up, as Dussel shows in his work from early to late. To develop a philosophy of liberation, one must take account of one's own position. As he claims, "A philosophy of liberation must always begin by presenting the historico-ideological genesis of what it attempts to think through, giving priority to its spatial, worldly setting."[11] Philosophy must understand the situated ground that is its condition—that is to say, the space from which it emerges. For philosophy of liberation, this involves considering its peripheral position within a larger world-system but also critiquing the position of the center. In modernity, the center of Europe forgets its own position: it claims unconditioned universality, which is predicated on such a forgetting, conflating its position with the neutral spaceless universal.

In *Philosophy of Liberation*, Dussel develops this sense of geopolitical-epistemological space, space as a battlefield, a conflictual field that has been historically and politically shaped. And it is within this space of conflict that thinking takes place and concepts emerge. The freedom of thought is still at play here, but the question is, what kind of thought emerges from this conflictual space in which it is enmeshed?

If space is a battlefield, thinking contributes to drawing up and retracing the map, either to tear down or reinforce its boundaries. This sense of space is operative in much of Dussel's work and captured in the phrase *geopolitics of knowledge*, which points to the necessary struggle of a philosophy of liberation to engage this battlefield and question the given divisions and boundaries of political spaces. It points to the entanglements of power and knowledge. Knowledge is not innocent nor is it unconditioned. When philosophy fails to question its situation and the distribution of space from which it emerges, it runs the risk of becoming a sedimented philosophy of the center. In an active sense, when thought engages its situated condition it can become a tool or method of struggle and resistance, and furthermore, it is able to think the limits and perhaps even the beyond of its own conditioning. This is the movement of thinking from within the present in order to think beyond it and to resist its perceived force of necessity.

There is another conception of space operative in *Philosophy of Liberation*, which emerges in relation to the first: ontological space. For Dussel, the geo-epistemology of space has ontological consequences (and vice versa). Those spaces that are known and grasped as central are reified as the spaces of being, while the outside, the exterior, or the periphery is taken as nonbeing. In this second sense, center and periphery are ontological concepts—that is to say, geopolitical concepts that have been ontologized. There is a violent conflation of a limited horizon of knowledge with a narrow totality of being. This spatial ontology holds that being is itself centered or central and nonbeing is peripheral. Phenomenological space is limited in this sense too, in that it involves only the space of those things that appear within a given system and would not grant existence to those things outside its field of vision.[12] However, beyond the purview of this system are the exterior and the excluded. The periphery, which does not count as full being, also cannot be and is not illuminated by the light of reason. Dussel succinctly claims, "The center is; the periphery is not."[13] What falls outside the system or the light of appearance is denied the plenitude of being, seen as non-existence or subexistence.

Thus, geopolitical space must be taken seriously, not as an abstract Newtonian reality or the framework of a neutral phenomenological subject engaging their lived experience, but as something that divides up reality itself with its territories and boundary lines.[14] Geopolitical space is something that places the subject in advance within a certain region of this struggle, as Dussel points out: "To be born at the North Pole or in Chiapas is not the same thing as to be born in New York City."[15] Geopolitical positions matter not just economically or socially but also epistemically and ontologically. It is not just the socioeconomic position of the peripheral subject that is in jeopardy but also the very status of their being and the practices of their knowing. The coloniality of modernity is not just about the uneven economic distribution and dispossession of wealth but also about the dispossession of knowledge and being, a process that takes place through the racialization of global (and local) spaces.

Since the conquest that generated the Americas and the debate between Juan Ginés de Sepúlveda and Bartolomé de Las Casas, the question that Europe as center has asked about peripheral subjects is whether they are fully human and capable of European (which is assumed to be universal) knowledge and religion. Already the very posing of this question, even if answered in the affirmative, reduces the (non)being of the periphery to the framework of the center. It establishes a caesura, the relation of nonrelation. The choices for the other are absolute exclusion (nonbeing) or assimilation, denying the otherness of one's being to be incorporated within the same. In both cases, the possibility of an alternative subject position is foreclosed. To be capable of religion means to be capable of accepting Christianity; to be capable of reason means adhering to and internalizing the tenets of European knowledge. Thus, it becomes clear how geopolitical space forcefully colonizes the concerns of ontology and epistemology. With the "discovery" of the periphery, the question quickly becomes what is the being of this other and does it fit within the ontology and system of the center.

There is, thus, an intersecting relation between these two senses of space in which the ontological variant is always entangled with a geopolitical history of struggle.[16] The geopolitics of the center and periphery lead to the ontology of the beings within this system who *are* and those outside who *are not*. Dussel points out that for Aristotle the Greek was human and that those who were not masters of their souls were slaves by nature, offering an ontological hierarchy of beings, which was, in turn, a justification for domination. Philosophy, when it develops from the perspective of zero-point hubris of the center, always runs the risk of offering such justifications of domination and creating a centered ontology in which what falls outside the system is cast as other or nonbeing.

Zero-Point Hubris

In Descartes's *ego cogito* we find the crystallization of the formula, which equates epistemology with ontology, certainty with being, while forgetting the space from which one thinks. The only being, that is, that exists, for the Cartesian subject is one's own thinking self. To establish its foundation, the subject centers on its own certainty and refuses to assent to the existence of any external being, the world, or its very own corporeality.[17] The attempt to return to the sensuous world will prove difficult after having discovered this foundation.[18] The threat of solipsism remains at the heart of this inheritance of modern philosophy, especially in the work of phenomenology from Husserl to Martin Heidegger, Jean-Paul Sartre, and Simone de Beauvoir.[19]

Epistemic certainty is derived from the only possible ontological certainty, and vice versa, as we see in Descartes's anchoring point, the "I think." "Archimedes used to demand just one firm and immovable point in order to shift the entire earth."[20] This immovable and yet placeless foundation is what Santiago Castro-Gómez calls the epistemology of zero-point hubris, an epistemology that must strip away the spatial and temporal conditions of its thought in order to develop its foundation as if it was in a vacuum or a no-place, the zero point.[21] The *ego cogito* can develop knowledge and be sure of its existence only from a position of absolute certainty and centering on its own ego. Yet in the same breath it must assume that the position from which it thinks is no place in particular. Because the mind is not an extended or embodied thing, the spatial position of the subject, its centering as a subject, is forgotten and instead assumed to be identical to the universal itself.[22] This is the movement through which a spatial ontology dominates the outside on the basis of its own self-centering, which is subsequently covered over. The particular gets taken for the universal.

Dussel's critique goes deeper to note that the ontological centering of the Cartesian subject is not simply an invention of Descartes's mind as he sits by the fire in an evening robe but rather one that emerges parallel to a geopolitical history of domination. This central ontology is a philosophical expression of the history of domination and the centering of Europe with respect to the rest of the globe after 1492: "Before the *ego cogito* there is *ego conquiro*; 'I conquer' is the practical foundation of 'I think.'"[23] The subject position of Cortés conquering the Aztecs in 1519 is the protohistory of Descartes's *ego cogito*: Cortés is the original embodiment of this subject position of absolute certainty, one that gained his self-certainty not through pure thought but through the conquest of the other. The "I conquer" subject claims an absolute right over the periphery. From an absolute standpoint, the "I conquer" destroys the other, does not recognize the other as a human being, and claims an absolute right over their land, body, and possessions.

Dussel's rereading of Descartes's *ego cogito* through its protohistory in the *ego conquiro* demonstrates the geo-critical nature of his philosophy, as it calls into question claims to universality resting on a centralized yet spaceless ontology. The Cartesian subject is connected to a history that stretches across the Atlantic, even if Descartes texts do not reference this connection to America in the ways that those of Locke or Vitoria do. In fact, the gap between the "I conquer" and the "I think" demonstrates the force of forgetting that is essential to the project of modernity: The violence that is built into the conquest of the other is forgotten and neutralized in favor of a subject position that appears to be free of any political or spatial engagements with the world.

Dussel develops the tools for a philosophical critique of modernity based on a spatial rereading of the history of philosophy in terms of the moments that were left out, forgotten, and covered over. The most important one here for Dussel is the forgetting of the spatial conquest of the Americas as the event where the possibility of modernity emerges. For Dussel, this critical moment of exposing the violent underbelly of modernity also points to the necessity of opening onto the subjugated knowledge of the other, developing a philosophy based on this exterior and the experience of exclusion. Thus, he refers to the philosophy of liberation as a barbarian philosophy since it will begin from the outside, from the peripheral space of nonbeing.

A Material Account of the Other

Dussel's turn to the other and spaces of exclusion has been the subject of a number of important and ongoing interventions and dialogues within Latin American philosophy, philosophy of liberation, and more recently decolonial theory. In terms of the spatial concepts operative in Dussel, several scholars argue that the center and periphery concepts inherited from dependency theory end up reproducing the very modern epistemic binaries that should be displaced.[24] This conversation has centered on the question of whether the peripheral other is turned into a new absolute, reproducing a form of totality in the wake of Dussel's critique of totalization.[25] Another concern emerges that pertains to reason and rationality in relation to the modern project: How can *liberatory* reason be salvaged, separated, or newly emergent against the *instrumental* reason of colonial violence? Can liberatory reason be affirmed and supported without an adequate appeal to the aesthetic, the imaginary, the material, and preconceptual sensibility? On these questions, Dussel's writings offer many rich pathways to emphasizing and valorizing the aesthetic, the material, and preconceptual sensibility once they are disentangled from the, at times, rationalist critique of postmodern thought and the metaphysico-theological reading of the other. In my account, transmodern thought incorporates a more capacious and liberatory notion of pluriversal reason alongside an embrace of the aesthetic and material conditions of liberatory thought.

In the 1970s, culminating with a book-length publication in 1983, the Argentine philosopher Horacio Cerutti Guldberg formulates one of the fiercest critiques of the philosophy of liberation and Dussel, in particular.[26] Cerruti's critique includes Dussel but applies to a much larger group and movement of liberation philosophers, mostly coming from Argentina.[27] Cerutti claims that philosophy of liberation is overdetermined by its emergence within the intellectual context of dependency theory and liberation theology. Dependency theory offered an account of the systemic nature of exclusion and poverty in the peripheral world that could be attributed no longer to a teleological narrative of the failure to modernize on the part of Latin America but instead to a history of domination and imperialism.[28] Cerutti claims that philosophers of liberation too readily accepted the claims developed by the economics and sociology of dependency along with liberation theology without developing their own account of systemic poverty and exclusion.[29] According to this view, liberation philosophy took the categories of poverty and exclusion developed in these other domains and used them as facts from which they could build their own philosophy without interrogating them further.

While Cerutti raises an important question here, such a critique fails to account for the profound epistemological transformation that took place through the social scientific discourse of dependency, which also shifted the terrain of philosophical debate in Latin America. For example, earlier philosophical discourses of positivism and modernization (discourses whose origin I would locate in the myth of modernity) had profound social, political, and philosophical effects in the shaping of Latin America.[30] Discursive struggle does not take place exclusively at a political, social, or philosophical level; instead these struggles overlap and influence one another: An earlier philosophical advance can later be translated into a sociopolitical one, just as a sociopolitical transformation of discourse may be translated into a philosophical innovation. In this sense, in the use of categories such as exclusion and poverty, what is at stake is not a question of reducing one level (say, the philosophical one) to the other (social, political, or economic) but of taking note of their resonating points of proliferation.

Epistemic breaks are not the exclusive power of philosophy. Cerutti clings to a modernist vision of philosophy as queen of the sciences in his criticisms, whereas philosophy of liberation grasps that multiple levels of discourse can be taken up philosophically not just as copy or epiphenomenon but to elaborate and exploit cracks in the dominant order of things.

A related critique of Dussel's philosophical architecture is raised, however, by Ofelia Schutte and Nelson Maldonado-Torres when they argue that there is a conflation of a geopolitical notion of "the other" with a metaphysical-theological notion of "the other."[31] Schutte's critique goes further to suggest that Dussel divinizes the other as beyond reproach and critique, essentializing and absolutizing the Latin American and the poor in problematic ways. Maldonado-Torres points out the Levinasian dimensions of his notion of other, emphasizing that Dussel's reading of Levinas's *Totality and Infinity* occurs at the same time as his discovery of dependency theory in 1969 (two years after his return to Latin America from his long sojourn in Europe). Although Dussel had long been interested in the systematic nature of poverty and the Cuban Revolution of 1959 had broken any possible consensus on the idea of "development" throughout Latin America, this theoretical confrontation of a political economy of dependency along with the metaphysics of exteriority in Levinas was formative for Dussel.[32]

Maldonado-Torres suggests that Dussel maps these Levinasian metaphysical categories onto the geopolitical notion of the excluded exterior and the impoverished other. However, Levinas's category of the other is a metaphysical notion that is supposed to defy all possible locality and empiricalness. If we could identify who or what the other is, they would cease to be absolutely other and run the risk of becoming categorized and identified within an epistemological or ontological system of the same. The other is, by definition, what defies the possibility of identity. In fact, the other haunts each and every one such that the foundation of identity itself is thrown into question. The other in this sense is not any one individual in particular but is the radical outside that calls out to each and every individual. The problem with Dussel's account, following Maldonado-Torres, is that he reduces the metaphysical other that is supposed to defy all specificity to a very specific empirical category: the marginalized poor or the excluded Latin American (a critique that echoes Schutte's earlier writings). At this point, philosophy of liberation runs the risk of imparting the absolutist and universalizing terms on the periphery that it precisely aimed to displace from the center.[33]

Instead of a need to abandon Dussel's philosophical architecture altogether, this critique points to the need of carrying out a materialist (rather than theological or metaphysical) reading of Dussel's concept of the other. An account that focuses on the terrain of struggle and the critique of modernity not through a pure outside but a materially constituted exterior. Rather than reading Dussel's other as absolute, as pure justice and pure disruption beyond reproach, it is more productive to read this as a material other that has been produced by a certain history. Dussel's account of ontology is itself sociohistorical and materialist. This social ontology shows how the European dominating subject has come to be identified with and in control of the totality and come to reign supreme over a positive sense of identity. Levinas is important to Dussel in making this discovery, but the influence does not require him to remain in the same metaphysical register. The radicality of these material claims are discounted by invoking absolute or (traditional) metaphysical categories, which Schutte, Maldonado-Torres, and Castro-Gómez are right to point out. Perhaps for this reason, in his later work *Ethics of Liberation in the Age of Globalization and Exclusion*, the role of the material and a materialist principle are situated at the core of Dussel's liberationist philosophy. The influence of Levinas's account of the other is still there in this work, but it is framed by the materialist reading and not the metaphysical or theological one.[34] This apparent material shift in Dussel's work from his 1998 opus, *Ethics of Liberation*, was already present in his earlier work as we see clearly in his materialist understanding of history, ontology, knowledge, and ethics.

Dussel offers a novel philosophical methodology, articulating a critique of colonial ontology and epistemology that exposes its force of domination. This social ontology echoes themes in Foucault's historical ontology of the present, in a way that avoids the mix-up between the level of concrete experience and metaphysics.[35] That is to say, Dussel demonstrates the practices and geopolitical history that have gone into the formation of ontology and epistemology rather than simply conflating the two. Furthermore, he demonstrates the geopolitical and practical effects that are exerted by ontology and epistemology. It is not a unidirectional flow from the practical to the theoretical, but instead these two levels contaminate, reinforce, and sometimes overturn one another.

Much of Latin American philosophy and liberation philosophy, especially, defy traditional disciplinary boundaries, in that they are engaged philosophies that do not abstract themselves from struggle.[36] Liberation philosophy emerges out of this tradition of discursive and material struggles in Latin America that occur within political, social, economic, and historical domains. Dussel finds it very important to remind his readers and his audience of this positionality in his philosophy: The positionality is itself an orientation to the inquiry.[37] Dussel's method is helpful in pointing to those places where new modes of thought are born out of struggle and out of an eclectic confrontation of discursive sources.

However, caution must be taken to avoid repeating the same absolute and universalizing claims that Dussel's very philosophy wishes to critique.[38]

The other should not become another absolute center to replace the European center; instead, the critique of modernity needs to aim for a decentering of all absolute positions. The notion of South-South dialogue points toward this decentering of all centers and instead opens up the possibility of a connection and set of relations that can be established between a number of different peripheries without center.[39] There is no absolute periphery to the center, but instead, the transmodern project seeks to articulate the relation of a plurality of peripheries. Instead of an exclusive and totalizing universality, transmodernity would be a global pluriversality, as the Zapatistas famously suggest, "a world in which many worlds fit."[40]

Toward a Transmodern Cartography

Even the Spanish conquests of old Mexico and Peru, which have been felt there like invasions from another planet—even those, irrationally for the Aztecs and Incas, rendered bloody assistance to the spread of bourgeois rational society, all the way to the conception of "one world" that is teleologically inherent in that society's principle.

—THEODOR ADORNO, *Negative Dialectics*

Modernity was born "when Europe could confront itself with 'the Other' and control it, defeat, violate it; when it could define itself as a discovering, conquering, colonizing 'ego' of the constitutive Alterity of the same modernity. In every way, this Other was not 'dis-covered' [*des-cubierto*] as Other, but rather 'covered-over' [*en-cubierto*] as 'the Same' that Europe had already been."[41] If 1492 or 1441 are birth dates of modernity, they also mark the origin of a very particular myth of sacrificial violence, alongside a process of covering over (*en-cubrimiento*) of the non-European, which was constitutive for Europe in its self-formation. Thus, the sedimented, excluded, or hidden other is carried forward into modernity; the other haunts modernity and forms its very condition.

In October 1992, the five-hundred-year anniversary of 1492, Dussel delivered a series of lectures in Frankfurt looking back at this birthdate of modernity, published as *The Invention of the Americas: Eclipse of "the Other" and the Myth of Modernity* (literally translated, "The covering-over (*el en-cubrimiento*) of the Other").[42] He takes up themes of earlier work on exclusion and silencing but offers a detailed philosophical history of how modernity is born in the fifteenth and sixteenth centuries by *covering over* and not simply *excluding* the other. He examines the process through which the other was covered over in the constitution of the Americas, while Europe was constructed as the unique site of universal truth. This is a history of the sedimentations of epistemic silencing and the ontological concealment of the other.

Modernity is constructed in response to an other, in confrontation with difference, but this other is simultaneously covered over, dominated, and silenced. Modernity is born, but along with it is born a myth of sacrificial violence done toward the other and an epistemic closure toward other subject positions. Dussel's aim in this work, ultimately, is to salvage emancipatory reason from modernity while separating it from its violent sacrificial myth. In this respect, he turns to the project of transmodernity as an overcoming of modern violence that aims to listen to the reason of the other and proliferate a plurality of epistemic possibilities and subject positions.[43]

Returning to a distinction I raised earlier, we can point out that two registers are at work in Dussel's understanding of the other, one material and one metaphysical.[44] The material reading shows the concrete nature of epistemic silencing and possibilities for excavating it from this sedimentation. If there are material subjectivities that have been racialized and invented as other, the transmodern proposal is to call not for the recovery of some pure other waiting to speak but for the opening of the epistemological field that violently forecloses the materiality of other voices. Thus, this covering over (*en-cubrimiento*) can be read as a material invention of the other rather than a metaphysical misrecognition, as Santiago Castro-Gómez suggests.[45] If the other was assimilated into the same without being *recognized* as other, the materialist reading shows that the voice of the other was silenced, the possibility of speaking and acting from an other subject position was *foreclosed*. The other was produced as a being that could be intelligible only from within a system of the same, of a given epistemic formation.

Given this diagnosis of the colonial cartography of modernity, we should turn also to Dussel's approach to redrawing the map in favor of a transmodern and pluriversal cartography. The colonial cartography emanates its shining universal out toward the brute and undeveloped periphery, while the transmodern map considers the planetary horizon of reason and the pluriversality of epistemologies and the plurality of locations from which they emerge. The transmodern approach needs to escape the binary and absolutizing logic of modernity that it seeks to critique. Here the transmodern project engages a decentered locus of enunciation to complete its critique of modernity. The center-periphery divide is still a binary product of modern reason that must be overcome through the transmodern project—it must be worked through and surpassed.

The macro-level critique of modernity's exclusion is a key moment of trans-modern critique, but it must be further pointed out that there is not simply one homogeneous center and one homogeneous periphery. Peripheries proliferate within the center, as do centers within the periphery.[46] Thus, the notion of transmodernity entails multiple decentered and wandering loci of enunciation (akin to a South-South dialogue) connecting horizontally, without the need for recognition from a totalizing center. The transmodern project envisions a cartography that moves beyond the binary dialectic between center and periphery.

From Postmodern to Transmodern

In *Ethics of Liberation in the Age of Globalization and Exclusion*, Dussel develops further some key notions from his 1992 Frankfurt lectures. He describes the corrective paradigm to Eurocentric modernity here as modernity from a "planetary horizon." In this sense, modernity was the first culture of *the center* of the first world-system.[47] Modernity is thus the product and process of the management of this centrality with respect to the periphery of the rest of the world. It is the result of the exertion of instrumental reason across the globe owing to its management of its central geopolitical position.[48] By understanding modernity from its "planetary horizon" we can see the conditions upon which Europe situated itself as the exclusive domain of pure thought and pure being, reducing the rest of the world to the barbarian, marginal exterior. This spatial colonization of center over periphery has dominated modernity and must be overcome with a transmodern project of liberation.

The corrective move to this colonization of thought and being is to shift the perspective to the underside of modernity, to that of the victims and the oppressed, the exterior or the periphery. The philosophy of liberation is then developed as a critical material affirmation of the life of the victims, and this requires a shift in the geography of reason and being from the perspective of the dominating (*ego conquiro*) to that of the dominated. Rather than a move to an absolute ethical other, this is a move to consider the planetary dimension of modernity and to include the reason of the other and the excluded spaces within, a pluriversal project of transmodernity. This account of transmodernity is not the pure metaphysical irruption of exteriority but rather the creative praxis of liberation that emerges from spaces of exclusion, silencing, or sedimentation.

In *Ethics of Liberation*, Dussel also undergoes a conceptual shift away from a more metaphysical register of the other in his earlier work toward a material concept of life and the destruction of the life of the victim.[49] Here he formulates a critical material ethics of life affirming the struggle of the victims against the systemic destruction of life in peripheral spaces, and the concept of the other is understood in terms of the victim suffering from the material and systemic destruction of life. Liberation is conceived as ethical and political praxis of the oppressed and those acting in solidarity instead of a theological or metaphysical irruption of the other. In this material sense, liberation and transmodernity refer to the praxis of the victims to articulate new forms of life and new epistemic positions within an era of global exclusion.

At the same time, in this shift to the material register of the other as victim, Dussel enacts the shift in describing his philosophy as transmodern rather than postmodern. Previously, Dussel embraced the label of the postmodern, as it would mark a movement past or break with modernity. In fact, he employs the term in 1977, before it becomes current in Continental European philosophy and prior to major publications such as Lyotard's *Postmodern Condition*.[50] In later works, he coins the term *transmodernity*, with the use of the prefix *trans-* to refer to a movement *across* and *beyond* modernity. Transmodernity breaks the stronghold of the center's exclusive claim to reason, shifting the perspective to recognize the reason of the periphery, a plurality of peripheries. Furthermore, this is a movement beyond modernity in the sense that the praxis of liberation from the periphery must seek to transform and not simply reform the given system of domination, which is one that is tied to modernity/coloniality. Transmodernity aims to overcome the provincialism and exclusion of Eurocentric modernity, with a global and dialogical reason. The shift from postmodern to transmodern aims to embrace and rescue the emancipatory possibilities of modern reason altogether, which postmodern thinkers reject.

If Eurocentrism has historically excluded and silenced the periphery, one of the tasks of transmodernity is to break the stronghold of this silence and interpellate the reason of the center to account for the reason of the other. According to Dussel, the periphery is not the *other of reason* but is the site of new claims to reason, the *reason of the other*. The reason of the other emanates from a marginal position from the perspective of a history of exclusion and inequality that cannot be accounted for within the communication community of the center. In conversation with the discourse ethics of Karl-Otto Apel and Jürgen Habermas, Dussel wishes to extend the boundaries of the communication community. Furthermore, he shows that one cannot think of the dialogical process of reason on neutral grounds but must account for the a priori exclusion of certain groups from this field. Dussel's point is to show that the other can interpellate the community of reason to demand inclusion and justice, to show that reason is not living up to its name when it continues to irrationally exclude.

In his critique of discourse ethics, Dussel accounts for the constitutive exclusion of those who are not included within the community of reason. This is an important intervention, and it opens up the borders that enclose the center. However, the movement of interpellation is still oriented within the cartography of a center-periphery relation. For the transmodern project to displace the violence of Eurocentric reason, it must radically displace the geography of this knowledge.[51] Without insisting on the autonomy of the praxis and epistemology of the exterior, one cedes too much ground to the modern bipolar cartography, without yet moving toward pluriversality.

Interpellation is the speech act in which the outsider erupts onto the scene of the center and makes a demand to hold them accountable. However, this demand risks becoming primarily a demand for inclusion and recognition if it does not, first of all, point to transformation. The other attempts to hold the center accountable for its supposedly universal values, but this does not dislocate the colonial geography of reason; it simply extends its borders. Liberation praxis would be required to speak the language of the master.[52]

These questions arise in relation to Dussel's engagement with discourse ethics and questions of communication arising out of Apel and Habermas. Another angle, through which Dussel carves out the meaning of transmodernity, is in dialogue with the dialectical critique of enlightenment in the early Frankfurt school of Adorno and Max Horkheimer. Here, Dussel suggests preserving the emancipatory pole of modernity while separating it from its violence to the periphery, the mythic dimension. Emancipatory reason can be *subsumed* (not Hegelian *aufhebung* but Marxian subsumption) in a planetary dimension that opens onto and accounts for the innocent other who has been victim to this mythic violence. By separating the violent mythic pole of modernity from its emancipatory rational pole, a global extension of emancipatory reason would cancel out its violent need for domination. However, the violence of the colonial emergence of modernity in a global battlefield suggests that reason and violence are entangled in a messier fashion. It is not simply a question of opening the borders of reason such that everyone can be heard. Reason itself is contaminated by its constitution in and through a violent exclusion and invention of the other, such that its very existence is predicated on this history of conquest. The language that reason speaks is deaf to the other, precisely because it was constituted through a destruction and domestication of the other's language.

To develop on this further, consider Dussel's third approach to transmodernity, in which he differentiates his transmodern thought from postmodern thought. This account arises in particular from his dialogue with Gianni Vattimo (which began after a trip to Turin in 1990). For Dussel, Vattimo embodies a kind of postmodern nihilistic thinking that offers no hope for a positive dimension to the overcoming or rescuing of modernity.[53] In Vattimo, Dussel sees the supposed "irrationalism" of postmodern thought, the notion that modern reason cannot be redeemed and is entirely reducible to violence and instrumentality or, worse, that nihilism leaves no hope for emancipatory possibilities. He rejects this notion that there is nothing that can be salvaged from the violent excesses of modernity and that reason is contaminated all the way down, but he still embraces much of Vattimo's work on the general postmodern critique of the excesses and failures of the modern project and the totalizing nature of a certain form of instrumental reason.

Transmodern thought is different from postmodern thought, then, on two fronts. First, the transmodern approach subsumes the critique of modernity under a planetary horizon and breaks the Eurocentric frame that sees modernity as only a European event and the critique of modernity as a primarily European problem. It is, in this sense, distinct from postmodern philosophy, which critiques the underbelly of modern reason but fails to critique its *colonial* underbelly. Second, the transmodern perspective is not against reason as such but instead aims only to separate the emancipatory aspect of modern reason from its violent myth. Both the first and the second aspects point to the affirmation of the epistemic position of the periphery or the "reason of the other."[54] Dussel claims that postmodern critique is, instead, against reason as such.[55]

The transmodern critique of the Eurocentric limitations of the postmodern highlights the failure to think the global dimensions of modern/colonial violence. The transmodern compels a rethinking and expansion of these critiques beyond the purview of the Eurocentric horizon.[56] One might consider the example of Foucault to raise the possibility of a dialogue. Despite the Eurocentrism of his archive, his spatial thinking offers methods that can extend beyond the content of his thought. If we are to fully understand the dimensions of modernity's underbelly, we must move to the planetary horizon of transmodernity as Dussel insists, yet postmodern critiques of domination may still provide invaluable methodological resources to further this critique. These resources offer possibilities for rethinking the question of how power operates spatially and how the history of Western reason relates to the spatial exertion of power and domination.

Dussel sees the postmodern critique of modern reason's excesses as an embrace of irrationalism. However, if reason is too quickly resuscitated or opposed against a binary of irrationalism, one runs the risk of foreclosing the creativity of thought in the exterior.[57]

Foucault's *History of Madness* might be critiqued as an example of praising irrationalism, yet if we look closer at this text, we see the complexities of trying to disentangle from a dominating monologic form of reason. In this sense, Foucault praises the creative and unrestrained power of madness prior to its enclosure by modern reason. He shows that there was an originary spatial division between those who were allowed to inhabit the space of the city and speak reasonably and those who were cast outside and silenced. The history of the great confinement that he traces is precisely a spatial enclosure and capture of a large population, isomorphic in character with the global totalization of colonialism after 1492. The establishment of the caesura that separates and consumes the other is evidenced in Foucault's history of madness just as it is in Dussel's history of the covering over of the other in America. In and through Foucault's account, it can be seen that the opposition between reason and unreason was consolidated through a violent spatial expulsion of a large group of people considered to be threats to the moral order of society: Reason could be mapped spatially just as was the Eurocentric reason of the center against the periphery. Foucault exposes the unsuspecting alliance between power and reason by showing that the violent expulsion and silencing of unreason would carve out the space in which reason could develop a monologue about madness, a medical discourse of mental illness. Thus, geopolitical conquest was intertwined with the emergence of psychiatric discourse about the human and the abnormal other.

Foucault aims to open up this stranglehold such that we might grasp what falls outside it and exists prior to the contamination of all epistemic positions by the discourse of a dominating reason. Here, what Foucault intends by reason is in line with Dussel's critique when he speaks of the violence of the myth of modernity that claims a right over some other in the name of epistemic superiority. For Foucault, this is the right of reason to conquer unreason, and it is a very specific form of dominating reason that for him emerged in western Europe in the middle of the seventeenth century with the separation of the mad, the poor, and the criminal into great spaces of confinement.

The stranglehold of modern reason would conceal its violence, forgetting the process that was built into the conquest of reason over madness, a process that was not accomplished through the pure objective gaze of scientific reason. Instead, the very subject position of this neutral gaze, the zero-point epistemology of the human sciences, was constructed on a history of domination. Foucault and Dussel (and Castro-Gómez) align in the aims of their projects, both of which demonstrate the spatial nature of epistemological exclusion, though Dussel is situating an earlier and more planetary epistemic silencing that occurs with the conquest of the Americas and invention of the other in the colonies.[58]

Foucault, in the first edition's preface to *The History of Madness*, dreams of recovering a space where a true dialogue between madness and reason might be had, rather than the nonconversation that dominates the contemporary psychiatric discourse on madness as mental illness. Psychiatry is a discourse about madness that never actually listened to the mad or understood the reason of the mad in their language but instead immediately silenced such a language and interpreted it from the already conquered position of mastery. Foucault writes, "There is no common language: or rather it no longer exists; the constitution of madness as mental illness, at the end of the eighteenth century, bears witness to a rupture in a dialogue, gives the separation as already enacted, and expels from memory all those imperfect words, of no fixed syntax, spoken falteringly, in which the exchange between madness and reason was carried out. The language of psychiatry, which is a monologue by reason *about* madness, could only have come into existence in such a silence."[59] Modern reason is incapable of listening to madness on its own terms. It can speak about madness only in a monologue of its own language, the scientific discourse of psychiatry.[60] For Foucault, this is an original division in the constitution of Western culture. Is there a possible way out of this monologue, beyond the romantic appeal to recover a lost origin? Foucault turns to the aesthetic and other modes of artistic expression that have escaped this monologue.

For Dussel, the original division that is forgotten in Western culture is the colonial constitution of modernity, a division left out by Foucault. Dussel operates from a global horizon of epistemic silencing and ontological erasure, the horizon in which the *damnés de la terres*, the "immense majority" of the earth, have been excluded ontologically and epistemically in a process that began over five hundred years ago.[61] It is not madness that Dussel wishes to excavate from the history of silence but the thought and practice of the Americas and peripheral spaces more generally, the reason of the other that has been covered over (*encubierto*) and materially excluded. Dussel's notion of transmodernity points to the opening of modernity to a pluriversal notion of reason, one that emanates from a variety of positions.

The reason of the other interpellates the reason of the center to include what it has excluded. Yet how would it be possible to reestablish the grounds on which a true dialogue might actually take place? Just as Foucault states that modern reason is capable of performing a monologue only about madness, a dialogue between center and periphery that does not disrupt the terms of the center is just another disguised monologue. Dussel is more optimistic that an emancipatory dimension of modernity can be salvaged from the wreckage of colonial violence.[62] Thinking this problem of reason spatially, we can say that the point is not for inclusion or incorporation of the periphery into the center but instead the establishment of networks of relation that extend horizontally in a plurality of directions.

The reason of the center has imposed its terms on the periphery for more than five hundred years. If this is true, the periphery must establish and wield its own grammar if it wishes to escape the stranglehold of this monologue about the other. This involves attention to new or excluded modes of reason along with practices of aesthetics and modes of organizing life and sensibility at preconceptual levels. The transmodern opening onto pluriversal reason must focus not only on the critique of reason as such but also on the very aesthetic, spatial, and bodily conditions of establishing alternative claims to and modes of reasoning. Furthermore, attention should be given to the creativity of the transmodern project in the tensions that it produces with respect to the hegemonic geography of knowledge, not just moments of possible dialogue with the center. There need to be meaningful and creative South-South dialogues on their own terms, not necessarily oriented by North-South dialogues.

The transmodern project is not, then, an antimodern project of reversal or the desire for some mythic purity of the outside. The claim is not to resuscitate some lost and covered-over origin point, as Foucault is also accused of in his preface to *The History of Mdness* and as Dussel is criticized for with respect to the metaphysico-theological other who is covered over. Instead, the transmodern project points toward the critical reactivation and engagement with subjugated knowledges, modes of being, thinking, and speaking that have been disqualified and silenced by colonial violence. This calls for the valorization of creative passions of the subject and the aesthetic dimensions of knowledge production, areas that may be traditionally cast out of the domain of reason by the totalizing center. Instead, the creative, relational, and aesthetic dimensions of thought production can give rise to a renewed and more capacious vision of reason, transmodern reason. The question that we have at hand, thus, is, how can a new grammar of reason be excavated and carved out as it struggles against the excesses of modern reason and attempts to establish a new sphere of intelligibility?[63]

Conclusion: Wandering Knowledge

> Knowledge is wandering much more than universal. . . . It proceeds literally from
> one locality to the next. . . . It is strengthened and liberated (intensely diversified)
> by such moving.
> —ÉDOUARD GLISSANT, *Philosophie de la relation*, my translation

Modernity is formed out of a battlefield of space, in which the reason of a centered subject and centered geopolitical position aim to reduce the globe to one
universal system. This occurs within the conflictual terrain of the first world-
system. In this process, Europe aims to expel the exterior and reduce the other
to silence, to conquer and to subjugate; making the other speak with one voice,
the voice of Christian European reason. The totalization is also split by a caesura,
separation and enclosure: It is in this relation of nonrelation in which the voice
of the other is reduced to unreason.

As Santiago Castro-Gómez claims, modernity is the process and the drive "to
submit the entire world to the absolute control of man under the steady guide of
knowledge."[64] With Dussel, I argue that this process first begins with the centering of Europe with respect to the rest of the globe in the emergence of the first
world-system and the ontological and epistemological silencing of the other. The
centering of Europe was not just a geopolitical phenomenon but also an epistemic and ontological one: It claimed that universal knowledge could emanate
only from one centered subject position, and it refused to recognize the validity
of other epistemic positions; similarly, it recognized the existence of only full
beings within the center within the domain of what was already known and appeared to the light of Europe, while everything outside this was cast into the
position of nonbeing or sub-being.

This account of modernity is meant to emphasize its spatial and colonial
matrix as a corrective to the standard accounts that remain blinded by the
temporalization of spaces as either primitive or developed. The spatial account also undermines the idea of a neutral subject that could be grounded
in a zero-point epistemology. As Nelson Maldonado-Torres articulates this
point with clarity, "Questions about space and the geopolitical relations
undermine the idea of a neutral epistemic subject whose reflections only respond to the strictures of the spaceless realm of the universal."[65] Dussel offers
a spatial account of modernity that destabilizes such a neutral epistemic position and brings us to affirm a decentered subjectivity. In this sense, I have followed the critiques that caution against absolutizing or reinscribing the other
as an unconditional subject.[66] Instead, I have argued here that the material
concept of the other, as invented and materially produced through a history

of ontological and epistemic subjugation, opens up an account that does not absolutize the other in this way.

This material notion of the other also works toward the way that the transmodern project is understood, not as the irruption of absolute otherness, but as a project to shift the geography of reason. The transmodern project is not only a global opening of reason but also a reorientation to the spatiality of reason itself. The sensory, bodily, imaginary, and aesthetic dimensions of the subject; their connection with reason; and the struggle against coloniality cannot be left behind in the shift to transmodernity. The pluriversality of multiple sites of enunciation for the reason of the other will also open attention to a more expansive notion of reason itself, liberatory reason. The transmodern project orients attention to global opening and also to the local specificity of different modes of orientation.

4. Archipelagos of Resistance

LIMITS OF THE MAP

The Antilles are an island bridge connecting, in "a certain way" [*de cierta manera*], North and South America. This geographical accident gives the entire area, including its continental foci, the character of an archipelago, that is, a discontinuous conjunction (of what?): unstable condensations, turbulences, whirlpools, clumps of bubbles, frayed seaweed, sunken galleons, crashing breakers, flying fish, seagull squawks, downpours, nighttime phosphorescences, eddies and pools, uncertain voyages of signification.—ANTONIO BENÍTEZ-ROJO, *The Repeating Island*

The coloniality of space functions both as an ordering and a nihilating regime, underpinning the architecture of modern systems of power and knowledge. In its nihilating dimension, spaces are constructed as empty, rendered void through the erasure of knowledges, lifeways, and spatialities of non-European subjects. The act of spatial negation operates as the very condition of possibility for the construction of a new colonial order. Transatlantic modernity thus gave rise to machines of silencing, destruction, and control. As I have argued, the production of order beyond the line does more than negate: it constructs the periphery as a zone where the distribution of bodies, knowledge, and lifeways are reassembled on a new grid. Yet there is more to the story of emptiness and destruction; there is more there than a simple and completed imposition of order.

The Caribbean archipelago offers a scattered entryway, an island bridge, to resistant landscapes and resistant modes of belonging across the Americas, beyond the invention of order. What the Cuban writer Benítez-Rojo describes as its "unstable condensations, turbulences, whirlpools" and its "uncertain voyages of signification" point to itineraries that escape the map of order and exceed the legacy of destruction. The discontinuous conjunction points to constellations and condensations that repeat, connect, and diffract across this meta-archipelago: a "certain way" (*de cierta manera*) of being organized, irreducible to the ordering of the colonial gaze.

The Caribbean is, after all, the intensive site where the colonization of the Americas begins, but there is something excessive about its landscape, something that exceeds what can or could be captured. Édouard Glissant describes this excess as a kind of openness and irruption of the real and the unreal and with an invented term, *irrué*:

> Every time I come back to the Americas, whether to an island like Martinique, the country of my birth, or the American continent, I am struck by the openness of this landscape. I call it "irrué"—a word I have invented, of course—implying irruption and rush [*ruade*], also eruption, perhaps a lot of the real and a lot of the unreal [*irréel*]. . . . In these spaces the eye cannot master the ruses and subtleties of perspective; one's gaze does not distinguish the vertical and the horizontal but just takes in a rocky piling up of the real. This American landscape that one finds in a small island or on the continent seems to me always equally "irrué". And that is probably the source of a feeling I have always had, of a sort of unity-diversity between on the one hand the Caribbean countries and, on the other, all those that make up the American continent. In this sense, the Caribbean has also always seemed to be a kind of preface to the continent.[1]

The Caribbean as preface to the continent of the Americas, or island bridge linking North and South; Glissant and Benítez-Rojo offer up images to think the unity-diversity between the Caribbean archipelago and the continent. They both have *irrué*, excessive, eruptive, open landscapes that give them an irreducible quality. From the expanse of the pampa in Argentina, the pre-Incan espaliered highlands of Chavín, Perú, the mountains and jungle of Mexico's southeast of Chiapas to the scattering of islands without center in the Antilles, the Americas, and the Caribbean are *irrué*.

In *Poetics of Relation*, Glissant describes how for the African peoples forcibly deported to the Caribbean, the Americas meant a complete uprooting and the expulsion into a threefold abyss: the belly of the slave boat, the infinite expanse of the Atlantic, and the uncertain birth of being thrown into an unknown new land.[2] In this sense, the Americas are also a space of abyssal uprooting. The abyss is what exceeds the totalizing gaze of the colonizer, whether it be the opacity of the wilderness or the sea. It is what is lost and cannot be recovered: not inscribed in the books of history but only in the subterranean landscape. And it is also what remains, as an insistent call of another reality speaking with a different tongue, not yet subsumed by the grid of modern/colonial reason.

Coloniality also evaporates Indigenous land through liquefaction, porosity, and abyssal logics of terra nullius emptying, as I show in chapter 2. Here also there is a resistant ground to this liquefaction, another sense of rootedness that refuses this colonial relation to land and sea. There is a world that remains, irrupts, exceeds, and even begins anew from the abyssal loss of coloniality.

Resistant ground, abyss, and the open eruptive landscape of the *irrué*: What kind of knowledge, what kind of resistance grows on this landscape that is not captured by order? Beyond or against the striations of modern order, are there forms of knowledge production and resistance flourishing on this *irrué* ground?

To situate this problem within the questions of spaces of order, control, and peripheralization raised in the previous chapters, this chapter echoes María Lugones's notion of resistance, that domination is never complete, that there is always another way of living and thinking that exceeds the totalizing map. When we have domination, when spaces are ordered in the name of subjugation, exclusion, or control, we will also find resistances. There are other modes of belonging beyond the totalizing map of power.[3] In this sense, I also return to the question of Indigenous resistance especially in refusing modes of state-based recognition that would seek to subsume Indigeneity within the orders of the coloniality of the state and its conception of territory.

Glissant's archipelago of resistance is not exclusive to the Caribbean, even if it is an exemplary site, or a preface to the Americas. The *irrué* landscape is taken up in resistances across the Americas. Glissant's turn to a horizontal model points to resonances with other actors and thinkers that see resistance in terms of a network that exceeds the trap of colonial recognition. Indigenous scholars Leanne Betasamosake Simpson and Glen Coulthard refer to this Indigenous relationality with the land that is irreducible to state- and capital-ordered private property as grounded normativity. The Indigenous land back movement is indeed an appeal to a certain root, but it is not the root of exclusive enclosure of private property or the colonial state form. It is instead an appeal to a relational connection with the land and its importance as the source of life in a wider ecosystem that includes the human as only one node in the network. It is a kind of place-based and relational form of resistance and resurgence. As Simpson writes, "We need to join together in a rebellion of love, persistence, commitment, and profound caring and create constellations of coresistance, working together toward a radical alternative present based on deep reciprocity and the gorgeous generative refusal of colonial recognition."[4]

Rather than attempt to understand colonization as a strictly top-down or dyadic hierarchy of colonizer and colonized, Indigenous Chickasaw scholar Jodi Byrd insists that we take a horizontal approach to the collision of colonialism, its cacophony.[5] She writes, "One such strategy is to read the cacophonies of colonialism as they are rather than to attempt to hierarchize them into coeval or causal order. . . . When the boundaries between worlds break down and the distinctive characteristics of each world begin to collapse upon and bleed into the others, possibilities for rejuvenation and destruction emerge to transform this world radically. The goal is to find balance."[6] Byrd describes the chaos and collision of cross-cultural understanding with the Choctaw term *haksuba*.[7] The collision of coloniality ruptures the boundaries between worlds, but the entanglements and cracks that emerge lead to possibilities for rejuvenation and transformation, especially when we don't see coloniality as a top-down and successfully completed project.[8]

Across the Caribbean and the Americas, we find the activation of new modes of subjectivity, communities of struggle, and practices of undoing constraining forms of order. The practice of domination is never total and never complete. In fact, sometimes its excesses, its most severe inflexibilities, produce newly mobile and creative modes of subjectivity through detours, plots to escape, modes of expression designed to decode and recode the master tongue, cross-cultural networks, or intensive relations between subjects confined.

In bringing the Caribbean archipelago into cacophonous dialogue with the American continent, we also confront the collision and entanglements among Black, Indigenous, and mestizo resistance. The construction of the coloniality of space through empty space or terra nullius functions through the dispossession of Indigenous land or the disciplinary governance of Indigenous and mestizo spaces. Yet the machine of coloniality and the invention of order is also built on the abyssal beginning of the transatlantic enterprise of the Middle Passage. In this sense, this chapter opens a dialogue between resistances that respond to the dispossession, confinement, and fungibility of Black spaces in the Caribbean and the dispossession, emptying, and ordering of Indigenous spaces discussed in previous chapters. There is a certain incommensurability between the coloniality of Black and Indigenous dispossession that I do not seek to conflate or resolve here; rather the aim is to show how the coloniality of space entangles and links these histories in discontinuous ways.[9] As Kris Sealey unpacks with respect to Glissant's notion of creolization as building opacities in relation, this position is both irreducible/incommensurable (opaque) and in relation, connected with others.[10]

Both land and sea have been found at the intersection of these emptying and ordering impulses in previous chapters, especially as we saw with the example of the traveler and the settler, the shifts between smooth and striated spaces. Following Glissant, we might say there is something irruptive or *irrué* about the landscape from land to sea and back: As Deleuze and Guattari show, both the sea and the mountain can be smooth spaces that exceed the capture of order. And the sea might be ordered in a sense by colonial rationality while land is rendered porous: The relation between these two is not fixed or opposed but rather articulated by the coloniality of space.

Tiffany Lethabo King develops the notion of the black shoal to describe the collision and cacophony where land and water meet. For King, this shoal also serves to unpack the entanglements between Black and Indigenous death and resistance. It is a meeting point where we see these spaces to be inextricable and incommensurable in their friction.[11]

The shoal is the space just offshore where land and water meet, where the smooth flows of colonial seafaring encounter obstacles and resistance: the threat of a rock piercing the boat from below or a pending moment of escape so close to land. This notion also provides a way of thinking how the dispossession and emptying of Native land meets with and brushes against the abyssal "liquid" construction of Black subjectivity.[12] King's notion works to show that binary constructions with Indigeneity associated only with land, on the one hand, and Blackness associated only with water and liquidity, on the other, fail to capture their points of mutual intersection and the places where these two racial histories cross from land to sea and vice versa in the Americas and the Caribbean. The shoal is not an anchor or totalizing pivot point to think these problems but both a foothold and a shearing point where *irrué* resistance pushes beyond facile constructions of either land or sea. The shoal is also the space of disruption: the shipwreck, the escape, the attack, or the threat of piracy. And we might read a certain irruptive and excessive quality into the collision of this space, something of the articulation of the *irrué* landscape of the Caribbean and the Americas. If the previous chapters primarily exposed the darker side of colonial modernity in terms of its violent and subjugating excesses, this chapter thus turns to the underside of modernity and the resistances that escape from, overturn, and create out of this experience. The heterarchy of colonial power leads also to a cacophony of resistant practices that are entangled with one another across the geographies of the Americas.

In closing, the notion of resistances that span and connect in unity-diversity across the Americas also returns us to the question of transmodernity, from Enrique Dussel. Transmodernity raises the possibility of thinking reason, resistance, and poetics adequate to global modernity that no longer forgets the centrality of the Caribbean and the Americas. Moving from the archipelago of the sea to networks of resistance built in the mountains or jungle, I consider the Indigenous struggle of the Zapatistas in Mexico as a transmodern example of building a world in which many worlds fit.

Archipelagic Thought and the Cry of the Plantation

> So history is spread out beneath this surface, from the mountains to the sea, from north to south, from the forest to the beaches. Maroon resistance and denial, entrenchment and endurance, the world beyond and dream.
> —ÉDOUARD GLISSANT, *Caribbean Discourse*

Glissant's archipelagic thought offers an account of a resistant mode of belonging in the Americas despite its abyssal beginnings.[13] Glissant shows that the history, geography, and imaginary of the Caribbean offer a model of thinking and acting opposed to the ordered, rooted, and linear models of European modernity. The ordered thought of modernity bounds itself to a gridded space and draws roots of filiation that enclose spaces, to hedge them off from others. This defines the bounded geography and geopoetics of what he calls continental (European) thinking.

Archipelagic thought is instead a geopoetics and geophilosophy springing from dispersion, uprooting, repetition, profusion of flows, and connections that form in a nonconstrained, chaotic, or dispersed space.[14] Underneath and beyond the invention of order, we find the formation of relation: dispersed archipelagoes connecting subjects, spaces, and languages across the Caribbean and the Americas. Another sense of modernity and another sense of space that challenges the coloniality of space is located in these spaces of relation.

The ordering project of colonial modernity could never be complete. At its foundation there is no unifying root; it is itself grounded on the abyss, which, unlike the colonial notion of emptiness, resists order. The Americas are the site for the heterotopia of order, a laboratory of power, yet they are also the birthplace of relation, a nonhierarchical and horizontal mode of forming connections and intensities with others. Glissant's reading of the Caribbean provides an alternative history of modernity, one that is built without totalizing roots but rather dispersed, rhizomatic roots, creating relations that spread out horizontally rather than vertically. Caribbean modernity is this history of nonhistory, the creation of thought and ways of life without singular roots or deep filiation. It is the creation of modes of thought and being from out of the horizontal collision of forces, languages, and practices of subjectivity that perhaps give rise to the first experiment of the creation of communities without filiation. This is, for Glissant, the more radical sense of modernity, which is not to say that modernity was not at the same time populated with the intensive project of ordering subjects and spaces but that it also gave rise to an underside populated with resistances and alternative modes of distributing space, knowledge, and subjectivity.

The Plantation was the defining institution of the Caribbean, according to Glissant and Benítez-Rojo, shaping the culture and society of the archipelago.[15] In this sense, we might speak of commonalities among Cuba, Martinique, the Dominican Republic, Haiti, and Jamaica that extend beyond linguistic and national boundaries. The history of colonialism, slavery, and the plantation in the Caribbean points to another dimension of the coloniality of space in the Americas at large, another set of techniques that is also at times coextensive with the techniques of ordering space outlined thus far. These Caribbean islands, such as Hispaniola (an island now divided in two by French and Spanish histories of colonization), were the first to be colonized by Columbus prior to becoming the site of nascent plantation societies: the preface to the Americas.

The Plantation was built on the abyssal subjectivity of diverse peoples uprooted from Africa and birthed onto the shores of the Caribbean alongside the abyssal emptying of the genocidal destruction and collapse of Indigenous societies. Although in Peru and Mexico the *mita* and *repartimiento* systems of Indigenous labor in the gold and silver mines would dominate alongside the spatial techniques of Indigenous *reducciones* and *congregaciones* discussed in chapter 1, the Caribbean was built particularly on the violent institution of slavery centered on the agrarian labor of the Plantation (especially sugar manufacture) rather than mining metals; yet as Benítez-Rojo shows, the machinery of all these institutions was entangled in the Caribbean.[16]

The Plantation defines the stultification and destructive subjugation of the Caribbean, yet Glissant takes this institution as privileged example that paradoxically gives rise to Caribbean modes of resistant relation. It is a space where uprooted peoples were firmly enclosed and ordered, yet this enclosure, this very rigid boundary, was also the principal failure and undoing of the Plantation. This rigid enclosure gave rise within deeply inhospitable grounds to a space of relation, an intensive formation of new communities and languages, a new knowledge, a new aesthetics. This underside of Caribbean modernity does not begin from subjective goodwill or the humanism of the modern project but instead shows how a deeply violent and destructive project gave rise to new modes of human relation precisely through its forcible and inhospitable cohabitation. This forced mixture of cultures and languages from various African, Indigenous, and European groups points to both the specificity and the universality of Caribbean modernity. It also shows that, on the other side of the line, there are resistant subjectivities in formation, not just nihilation and destruction.

The islands of the Caribbean are divided by the sea, populated with diverse peoples, speaking Dutch, English, French, Spanish, and many Creoles drawing on the influence of Indigenous and African languages. However, these islands and this sea share a history of uprooting and violence, and a history of what Glissant calls creolization. Creolization refers to the collision of cultures, tongues, ideas, peoples, and struggles in the Caribbean, along with the creation and poetics that result from such a process. The unfolding of creolization is always unpredictable, and it is especially located in the linguistic problem of how to speak a common language when common ground is lacking.

The linguistic unity underlying the study of the Hispanic Americas or the Spanish Caribbean is thus also disrupted by Glissant's articulation of another mode of relation that connects the people of the Caribbean despite their uprooting. Glissant insists that there is a language or mode of expression (*langage*) of the Caribbean that exceeds the particular grammar of individual tongues (*langues*).[17] The work of the Cuban writer Alejo Carpentier written in Spanish might share something of this *langage* with the Barbadian poet Kamau Brathwaite, whose works are in English, or Glissant, who writes in French and whose native language is Creole. There is a link that ties these modes of expression together, which is not readily visible on the surface. It is a relationality that cuts underground or underwater. As Glissant quotes from Brathwaite, in several places, "The unity is submarine."[18]

Glissant thinks from the experience of uprooting without return: the creation of Caribbean thought and culture from its groundless ground and its subterranean connections. He thinks an abyssal beginning, a futural projection, or a becoming of Caribbean-ness rather than a recovery of a rooted past. He will locate this unity in a perpetual process of becoming, collisions, and flows: a process he refers to as the "poetics of relation" but also relatedly as creolization.

Glissant's focus is the specific experience of the transatlantic slave trade and the uprooting of diasporic African peoples into the Caribbean. This is an irreducible experience of memory and suffering; it is not open for all to understand or relate with. It is an experience that cannot be rendered transparent, known, or grasped. Thus, Glissant proclaims famously a right to opacity. A right to not yield to objective transparency, such that anyone from any perspective could grasp and know you as if an object. The right to opacity points to the nonsubsumable nature of particular subjects and their history.

Yet the claim to opacity is not a movement toward self-enclosure and isolation. It is not understood in opposition to relation but as perhaps a precondition to the creative and vitalizing practice of relation that draws up connections across difference. Irreducibility and difference must be held together with a network of relations. In this, Glissant also conceives of Caribbean dialogue and the reason of Caribbean poetics as one that gives-on-and-with (*donner-avec*), making connections, forming relations, without the requirement for subsumption. One can relate, dialogue, reason with the other without the need to take up and grasp transparently. Here he makes note of the root of the French term for "understanding," *com-prendre*, literally "grasping-with." Instead of grasping the other, opacity gives-on-and-with the other. *Donner-avec* instead of *com-prendre*. Caribbean reason connects without containing.

From the specificity of these new modes of relation born in the Caribbean, Glissant aims to think a nontotalizing totality, a kind of universality that he calls the *tout-monde,* or all-world. Caribbean creolization is exemplary of a shift in the whole-world, the creolization that confronts all cultures. The meta-archipelago of the Caribbean is paradoxically at the center of this history, but it is also one that diffracts across the whole-world and is found to have its own drama and history in various locations.

Language is at the heart of this understanding. The articulation of life, suffering, resistance into language is at stake: the expression of a new language of the Caribbean that arrives through detours and creolization. It is not any specific tongue (*langue*) that is spoken but the mode of expression or language (*langage*) of the Caribbean. As Jean Bernabé, Patrick Chamoiseau, and Raphaël Confiant write in their *Éloge de la Creolité*, "Creoleness is not monolingual. Nor is it multilingualism divided into isolated compartments. Its field is language. Its appetite: all the languages of the world. The interaction of many languages (the points where they meet and relate) is a polysonic vertigo. There, a single word is worth many."[19] A specific kind of universality emerges from this expression of language in the Caribbean: the connection with the other, with many others, with the *tout-monde*. Language as the formation of relation, rather than an enclosed community: This language escapes the confines of the Plantation that sought to enclose and instead proliferates across the Caribbean and the world, despite the variety of tongues.

Glissant describes the orality and expression of the Plantation, the transposition from the suffering cry to the language of the world: "This was the cry of the Plantation, transfigured into the speech of the world. . . . The place was closed, but the word derived from it remains open."[20] Further, he adds, "this is the only sort of universality there is: when, from a specific enclosure, the deepest voice cries out."[21] There is no abstract universality, no universality that is not grounded from a particular enclosure or spatial distribution, but there is the universality of this deepest voice: the openness that explodes from within the most rigidly enclosed space, the network of connections and relation that emanates from this place. In this sense, Glissant is a thinker of pluriversal universality, a radical project of relational and resistant knowledge that forms a network across disparate spaces in the Caribbean, the Americas, the Atlantic, and across the globe. He rejects the notion of simply being a nationalist or identitarian thinker but instead insists on the open and relational dimension of identity, while the right to opacity ensures that this identity is not swallowed up by the colonizing rootstock of the center.

In this way, the poetics of relation develops, excavates, and connects a form of relation that resists the domination of center by the periphery. As opposed to the model of roots and trees that has dominated Western thought, Glissant deploys a creolized notion of the rhizome as a model of relation without deep filiation.[22] Arborescent roots are vertically organized, extending down into the soil to the very solid origin of things; from there they protrude upward, branching out into the perfect binarization of knowledge. All things have their principle in one origin that can be traced back and followed out through a linear history with binarized branches. This is not just an epistemological image but also a political one that can be tied to the image of the state as an apparatus that draws out the roots of people and attempts to bind groups around sedentary principles of rootedness. Instead of the binding force of stagnation or an ontological claim about being, rhizomes are networks of becoming and alliance. They are continually in flux and in motion while also forming intensive connections and alliances with others. As Deleuze and Guattari write, "The tree is filiation, but the rhizome is alliance, uniquely alliance."[23] For Glissant, this is precisely the case for Caribbean reality as it operates without filiation. It is not tied down to an enclosed root origin or a linear history but instead operates on principles of alliance and relation.

In this way, he critiques a certain notion of rootedness, what he calls the totalitarian rootstock, that aims to reduce everything to the same. The rootstock constricts and consumes every outside within one linear and vertical model. As the roots extend, they produce homogeneous, striated space that can be reproduced, charted, and controlled anywhere. In the history of colonization, the totalizing root impulse aims to spread the uniformity of this striated network from the center of Europe across the globe and extend its roots into each and every other.[24]

Glissant sees that the geography and geopoetics of the Caribbean are populated by a rhizomatic enmeshed network of roots that spread out across a series of diffractions and dispersions in a horizontal way. This understanding does not reject root structures as such, as they are not necessarily suffocating. The rhizome is itself an "enmeshed root-system, a network spreading either in the ground or in the air, with no predatory rootstock taking over permanently. . . . The rhizome maintains, therefore, the idea of rootedness but challenges that of a totalitarian root."[25] The rhizome is a way of thinking the reality and the resistances of the Caribbean, it is a way of thinking something that escapes or overturns the totalizing violence of the modern rooting project. Caribbean modernity is rhizomatic, pushing beyond Deleuze and Guattari's formulation, which almost liquefies without any specific enmeshed network, unlike the entanglement of the Caribbean mangrove. Glissant is not against roots and belonging but against those who seek to close off all relations. Glissant offers a historical ontology of the Caribbean: He aims to think the reality of the present in terms of how it has been violently shaped and constrained, but he also finds a brilliant creative project that is oriented toward the resistances that link up in relation.

Territory and the Self: Land Beyond Territory

Glissant's history of the present is evidenced in his account of the formation of roots and root structures in relationship to the history of colonialism, errantry, exile, nomadism, and state formation. The totalizing rootstock becomes most firmly established in the world during the history of colonialism along with the concomitant formations of the Westphalian state.[26] During this period there is an immense growth of the notion that identity comes from the relation between the territory and the self in opposition to an other that is kept out of bounds. The self is defined in terms of one's relationship to a defined territory and the lineage of culture within that territory, and it grows out of the soil of the mother- or fatherland. This has not been the case for the history of most great civilizations and communities but is, instead, quite unique to modern Western history. Thus, as Glissant writes, there is "an immense paradox"—namely, that "the great founding books of communities, the Old Testament, the *Iliad*, the *Odyssey*, the *Chansons de Geste*, the Icelandic *Sagas*, the *Aeneid*, or the African epics, were all books about exile and often about errantry."[27] The traveler who becomes the settler or the traveler who was always already desiring settlement is unique to the modern Western project.

The history of colonization, of course, involves its own history of voyage and exile. The age of exploration was profoundly transformative for Europe and the sense of European identity. As I have argued here, this voyage and the construction of the Americas as an empty space, the conquest of the other, and the heterotopia of constructing a new space were at the heart of European claims to universality. Glissant describes this time period as one in which identity was consolidated by a relationship between "voyage and the other." During periods of conquest and the establishment of empires, we see a movement in which "self-definition first appears in the guise of personal adventure" and "conquerors are the moving, transient root of their people."[28] The "I conquer" (*ego conquiro*) form of subjectivity, described by Dussel, gets formulated in this period when a community defines itself in relation to an other that it conquers and through a heroic voyaging subjectivity that it proclaims for itself. We might be tempted to think, then, that colonialism is not about a history of planting roots but only one of discovery and voyage.

Glissant is quick to point out, however, that conquest and colonial voyaging reveals itself as a "devastating desire for settlement." The West is the process and the project through which this movement becomes fixed.[29] The intolerant root begins to take shape, and the concept of a people rooted in a defined territory from which they receive their identity in opposition to others therein takes shape. Colonization and conquest will no longer involve the perpetual movement of the root but the instantiation of the center and its predatory rootstock spreading and settling in the periphery.[30] Locke's understanding of property as enclosed settlement on an empty wasteland starkly exemplifies this logic. For Locke the English colonial imaginary of a new, empty space afar is essential to the notion of a space that can be cultivated, enclosed, and settled: a space where land can become part of the self, the propriety and property of the self that is extended and formed through labor.

To differentiate violent movement directed toward settlement and a notion of errantry or movement that goes against the root, Glissant defines two types of nomadism. There is arrowlike nomadism, which describes this nomadic movement of conquest and exploration; yet it is movement that ultimately seeks to transplant roots or to plant roots in a newly conquered space, as the Spanish grid in the New World or the English hedge in the "wilderness." This nomadic movement is contrasted with circular nomadism, which moves from place to place without having any rooted relation to a particular territory. Circular nomadism does not aim to establish filiation in a territory; instead the bonds of the nomadic group are established through horizontal relation.

Glissant proposes that the Caribbean is not consumed by this intolerant and predatory rootstock but is instead a space of "an enmeshed root system," a rhizomatic root system in which relation spreads out horizontally. He turns, thus, to "the thinking of errantry and totality."[31] Errantry does not establish identity through filiation within a territory. It is a mode of thought that "silently emerges from the destructuring of compact national entities that yesterday were still triumphant and, at the same time, from difficult, uncertain births of new forms of identity that call to us."[32] Errantry opposes the stagnant totalitarianism of roots that tie down the territory and the earth to a rigid notion of identity. Errantry proposes instead a *relation* with the other: what Glissant calls "totality" refers to a relation with what is outside us and what stands in this horizontal relation beyond a particular territory. "In this context uprooting can work toward identity, and exile can be seen as beneficial, when these are experienced as a search for the Other."[33] In this sense, the imaginary of totality "allows the detours that lead away from anything totalitarian."[34] Like Dussel's critique of totalization, Glissant's search for identity in exile does not allow subsumption of the other as if an object. Instead, it involves the search for the other, the formation of relation that still allows the space for opacity; giving-on-and-with rather than grasping transparently. Thus, this form of relation does not constrain and suffocate the becoming of the other. It is a relation that connects with networks of difference in search of the other rather than shackling and grafting the other onto a preestablished root. The rhizome exists on a smooth space without hierarchies and without the subordination of the abstract grid of striation. The rhizome creates open points of connection and intensity. It indicates possible breakages in the shackling of modern coloniality, which subsumes the other under the categories of the same.

For Glissant, this is not a disavowal of all roots or identity but instead directs us to a root without center and an identity without enclosure. Rather than the processes of colonization, which forced colonized people to establish an identity in opposition to the roots of the center (through various national projects)—thus reproducing the devastating dangers of the totalizing rootstock within their own histories—the Caribbean practice of decolonization as relation moves beyond this limit. Glissant claims that this is a movement toward a form of identity that feels the earth as a smooth surface and land as a passageway toward a possible relation.

Glissant writes in this regard, "Relation identity . . . does not think of a land as a territory from which to project toward other territories but as a place where one gives-on-and-with rather than grasps."[35] Root identity is expansionist and totalizing. All things have their principle in one origin that can be traced back and followed out through a linear history with binarized branches. Glissant's notion of giving-on-and-with (*donner-avec*) is supposed to be instead a way of rooting oneself in space without becoming fixed in a stable uncontaminated identity. Relation identity as an enmeshed root system that connects with others across space rather than an enclosed or expanding model.

Resistance Beyond the Petrification of Enclosure

This voyage "in a certain kind of way," from which one always returns—as in dreams—with the uncertainty of not having lived the past but an immemorial present . . .

—ANTONIO BENÍTEZ-ROJO, *Repeating Island*

Here, memory belongs only to objects. . . .
In our countries victimized by History
where the histories of many peoples
are intertwined, works of nature
are the true historical monuments.
—ÉDOUARD GLISSANT, *Faulkner, Mississippi*

To return to the question of resistance, it is worth asking how Glissant offers not just an account of Caribbean reality but a poetics of resistance. In one register, Glissant seems to simply describe the Caribbean: It is an archipelagic space of relation and errantry. Yet it would be a mistake to see his account as a simple description. The imaginary and the poetic aspect of memory and creation are crucial for Glissant. Thought is neither static nor objective but involved in a process of transformation and becoming. He conjures from the imaginary of the Caribbean and creates a counterhistory out of its abyssal and unknowing past that he refers to as nonhistory: Telling history from the underside of modernity and from the side of those who were denied a voice, denied writing, denied a historical record involves an appeal to the aesthetic and the imaginary.[36] He writes, "Thought draws the imaginary of the past: a knowledge becoming."[37] Or as we can see from the epigraph to this section, one must trace out a memory from the landscapes and monuments that remain—the silent objects, the sea foam, the mountains, or the forest. This way of tracing history from an abyssal and *irrué* ground is also about destroying the perceived stagnancy of the present and locating history from the subterranean to the sea foam, to the mountain as it motivates the possibility of another way of being and acting.

Glissant draws from the history of the Caribbean, dispersed on the surface but unified in a submarine fashion that cuts below the surface. His central examples of relation in the Caribbean are both violent and traumatic sites of memory: the slave ship and the Plantation. These are also sites of nonmemory and nonhistory in the sense that there is very little record left from which to read and decipher these histories. The slave ship, or the open boat, is an abyssal image at multiple levels: There is not only no adequate mode of representation to formulate the history of this trauma but also no adequate way to account for what these experiences of violent uprooting could or did mean to the people who were birthed onto the shores of the Caribbean or buried in the abyss of the sea.

Yet knowledge is born of this experience of the abyss, that a new foundation without foundation will serve for the birth of a Caribbean mode of thought and being. He writes, "Thus, the absolute unknown, projected by the abyss and bearing into eternity the womb abyss and the infinite abyss, in the end became knowledge [*connaisance*]."[38] What is the form of knowledge that Glissant proclaims here; how does something or someone emerge from the abyss and absolute unknown to become knowledge? How can we locate this resistant sense of modernity in the formation of a knowledge shaped out of the *irrué* and the abyss rather than order and enclosed ground? Glissant's project is one of resistance, poetic and political, in that he does not take this history of destruction to be simply a traumatic wound that binds the violent and irrecoverable roots of the past but to be beyond that, a projection into the future, into a Caribbean becoming.

Glissant, in this way, offers a singular reading of the Plantation as a site of relation and creolization. The Plantation is a space of domination that populated modernity across the Americas, especially the islands of the Caribbean, the Southern United States, Brazil, and the Caribbean coast of Latin America. It is a site where we find an intensive concentration of an impulse for order, enclosure, and domination. Yet, as Glissant argues, it is also a site in which some of the most intensive mixtures of language, thought, and relation take place, giving rise to the formation of communities without roots or claims to filiation. In fact, Glissant sees in the Plantation a place that gave birth to creolization and relation in the modern world.

The Plantation is "one of the bellies of the world," whose "boundary, its structural weakness, becomes our advantage."[39] The language of the Plantation was formed through detours and escapes; it was formed not just as the transposition of language from Africa, Indigenous peoples, or Europeans to the Caribbean but as the creation of something new. The nonhistory of the Caribbean is told as counterhistory against the imposing totalizing history of the West. The orality of the voice exceeds the structural boundary of the Plantation, even if it fades into the wind without being recorded on the page. Caribbean reason is about this mode of creation that establishes a language of the world through repetition and detours, or what Benítez-Rojo calls the "repeating island."

In chapter 1, we saw that the order of colonial cities along with a grid-like structure provided a model for knowledge formation and a disciplinary method for the subjugation of Indigenous subjects. The Plantation shares this intensive ordering impulse and the heterotopic and disciplinary structures of the colony. It shares the impulse to perfect and control a space that we saw in the gridded urbanism of Spanish colonial cities, though its techniques of enclosure are more violent and severe.

Glissant shifts the vantage to the underside of this technique. The Plantation is a technology of enclosure and constraint and is supposed to operate as a self-sufficient autarky without opening. There is to be no escape, no voice or body that leaves the Plantation and its stifling culture of violence and subjugation to agrarian work. Yet this desire for sealed enclosure cannot shut down the process of relation, the contagious connection between subjects, languages, and the outside. In one small space so much domination is concentrated against a group of people, and yet these subjects give rise to new modes of expression, detours, creolization. As Glissant writes of these many contradictions, "How could a series of autarkies, from one end to the other of the areas involved, from Louisiana to Martinique to Réunion, be capable of kinship?"[40]

A technique for subjugation was developed, one intended to close down and shut off the world, yet paradoxically, this technique also gave birth to what Glissant calls relation. Similar cultural expressions, similar ways of speaking (*langage*), similar modes of resistance resonated among these dispersed places that were supposed to be closed to the world. The creole language, jazz, blues, calypsos, salsas, reggae, storytelling, and literature transformed "and assembled everything blunt and direct, painfully stifled, and patiently differed into this varied speech."[41] The cry transformed into speech is the poetic movement that Glissant traces out of the Caribbean, the knowledge that it produces. Despite the dispersion of different particular languages and different colonial inheritances, where a Martinican might feel a stronger connection to France and a Puerto Rican to the United States or Spain rather than to fellow Caribbean nations, Glissant finds that there is a common language (*langage*) or form of expression of the Caribbean that unites in dispersion.

The Plantation is a space of concentration of domination and order in an intensive way that differs from and echoes the Indigenous enclosures of the *congregaciones* and *reducciones* in certain ways. It is marked by the will to organize and extract forces from bodies yet also marked by the excess of fungibility and its abyssal logic that cannot be reduced to economic terms. Yet its excessive principle of petrification and constraint is also a principle of its failure. It fails to produce cellular subjectivity. The Plantation does not strictly divide subjects into their own cellular space and cellular subjective isolation but instead forcibly places them in a space of a cohabitation, inhospitable cosmopolitanism. In this sense, it gives rise to new modes of forming intensive connections with others, new modes of building community and resistance, and new ways of speaking and developing one's own grammar or language (*langage*) to take detours around the tongue (*langue*) of the master.

Aesthetics of Relation and Transmodernity

Relation is the knowledge in motion of beings, which risks the being of the world.
—ÉDOUARD GLISSANT, *Poetics of Relation*

We shall live its discomfort as a mystery to be accepted and elucidated, a task to be accomplished and an edifice to be inhabited, a ferment for the imagination and a challenge for the imagination.
—JEAN BERNABÉ, PATRICK CHAMOISEAU, AND RAPHAËL CONFIANT, *Éloge de la Créolité*

Glissant is a thinker of the geography and history of the poetics of thought, its sedimentations, scatterings, and memory fragments. As we saw in chapter 3 with Dussel, for him universality ought not be thought from a neutral spaceless position. Dussel's notion of transmodernity exposes the Eurocentrism of the pretension to a spaceless universality that just happens to be coincident with the particular history of European space. Transmodernity intends to think and open onto the pluriversality of global thought. Glissant argues a similar line insofar as he insists on the situated nature of thought. Reflections on thought usually retreat to an area without dimensions, "but thought in reality spaces itself out in the world. It informs the imaginary of peoples, their varied poetics, which it then transforms, meaning, in them its risk becomes realized."[42] Thought is lived in the world, and its risk is embodied and enacted in a people. Thought is engendered in the space of the world in relationships of peoples and the poetics they use to grapple with the world. It is not born outside the world and subsequently spaced out therein; instead, it is a process immanent to the world. For Glissant, there is no realm of ideas separate from spatial distribution, fossilization, flows, spirals, and consolidations. His philosophical task, in this sense, takes up the geography of ideas, the history of their sedimentations in the landscapes and forgotten objects that scatter the Caribbean, along with an understanding of the power relations that they consolidate. His work is oriented toward resistances through the creation of new modes of thought and new relations in the present.

It might be asked, then, what is the importance of the aesthetic in think-ing modes of resistance to colonial forms of knowledge/power? The aesthetic question of resistance returns us to the divisions of the world implanted by modernity/coloniality, the battlefield of space, which was also the dividing line between different forms of knowledge. For Dussel, the response to the prob-lems of modernity's excesses is the dialogical upsurge of the voice and reason of the other. Yet I have argued that it is important to develop a new grammar or language (*langage*) that would not be indexed back to the tongue (*langue*) of the master and a way of talking about reason that would not be measured by the master's measuring stick; to instead be more attentive to the supposedly irratio-nal dimensions of thought and action that are too easily dismissed by colonial reason, along with the creative passions that burst forth in liberation struggles.

In this sense, Glissant provides insight into another pathway of thinking the question of transmodernity. Transmodernity proposes an orientation toward the global dimensions of modernity and the pluriversality of reason and the practice of the reason of the other that extends across varied global spaces. Poetics of rela-tion carries this thought by considering the nature of intensive connections that extend between disparate subjects across an archipelago of locations. Glissant's aesthetic mode of thought allows us to think a form of rationality that is not strictly divided among the poetic, the irrational, the creative, and the imaginary. Instead, the imaginary and the creative are crucial tools to the production of a *reason* of the Caribbean and the Americas, more generally.

Glissant's understanding of relation and the possibility of Caribbean reason and dialogue refuses the notion of transparency. He thinks against the notion of fully grasping or knowing any particular subject, the need to give an account that would clarify without remainder. On this note, he famously proclaims "a right to opacity," a right not to be grasped or thoroughly known.[43] Relation is not strictly a form of rationality or a dialogical mode of being but is instead what he describes as a mode of giving-on-and-with (*donner-avec*) instead of grasping or taking up (*com-prendre*).[44] If we take Glissant's more aesthetic image of thought, transmodern rationality would be a method of giving-on-and-with rather than attempting to grasp or render transparent. The right to opacity points to a proc-lamation of nonintelligibility and nonrepresentability that is constitutive of a creolized subject. Glissant is not proposing Caribbean nationalism; he does not seek isolation or a return to some newfound enclosure. He seeks to articulate a form of relation that would connect without being subsumed by the other, a network that doesn't get taken over by the colonizing rootstock. Dussel and Glissant find a point of transmodern resonance on the question of the voice and the cry. Dussel raises the question of transmodern reason as an issue of listening to the reason of other. The question is whether this involves an interpellation of the listener or the demand for a transformation of the listener or whether there are additionally transformative modes of speech and transformative modes of the cry. The cry of the plantation is not yet speech and not yet language, yet Glissant points out that it gets formulated into a language of this world through a detour: through the creolization of jazz, among other expressions. Dussel's thought also points to the transformative character of the voice and the material cry of the victim or the excluded one who aims to transform the injustice of the system. For Dussel, the cry of suffering is the moment of transition from material life to the linguistic and eventually rational community.[45] Glissant shows us that the cry is itself already language (*langage*), a fundamental poetics of the Caribbean that cannot be rendered transparent yet offers a creative mode of expression and reason. The material struggle to create horizontal spaces of dialogue can avoid the threat of instrumental dominating reason if it avoids the risk of rendering the other transparent. Perhaps it is the spiral-like movement of the decentering of the subject that moves outward and disseminates its methods of resistance outwardly rather than concentrating them inwardly in one secure place that could be mastered, the poiesis of reason and resistance.

The transmodern cry can also be heard in the echoes that have spiraled out of the Zapatista mountains of Chiapas as a *caracol* (the snail that symbolizes this Zapatista slow spiral of relationality). The Indigenous rebellion that was declared there in 1994 continues to this day and launches into the imagination of a new future. Their vision is a hyperlocal one of creating autonomous zones of governance, education, and daily life beyond the orders of the Mexican state, but it is also a transnational and international struggle that sees its battle echoed by and connected with many other sites across the globe. As they declared at an International Encuentro for Humanity and Against Neoliberalism in 1996, "The word born within these mountains, these Zapatista mountains, found the ears of those who could listen, care for and launch it anew, so that it might travel far away and circle the world. The sheer lunacy of the calling to the five continents to reflect clearly on our past, our present, and our future, found that it wasn't alone in its delirium. Soon lunacies from the whole planet began to work on bringing the dream to rest in La Realidad, to bathe it in the mud, to grow it in the rain, to moisten it in the sun, speak it with each other, to bring it forth, giving it shape and substance."[46] The cry of these mountains, the delirium and lunacy of its dream, begins to take shape in the mud, rain, and sun as it spirals out across the world. This transmodern image shows how the formation of another world circulates between transnational solidarity and local opacities of specific struggles.

The image of contagious resistance is drawn out further in the Fourth Declaration of the Lacandon Jungle with a sense of this transmodern relation that is launched forward first by the word or the cry, "Our blood and our word have lit a small fire in the mountain and we walk a path against the house of money and the powerful. Brothers and sisters of other races and languages, of other colors, but with the same heart now protect our light and in it they drink of the same fire. The powerful came to extinguish us with its violent wind, but our light grew in other lights. The rich dream still about extinguishing the first light. It is useless, there are now too many lights and they have all become the first."[47] The world in which many worlds fit cannot be extinguished, because it has found these points of resonance, relation, horizontal connection that light up many lands.

The Zapatista rebellion speaks many languages, but it has found a common language too, what Glissant would call a *langage*. Dussel dedicates *Ethics of Liberation in the Age of Globalization and Exclusion* to the ethical path opened in the Zapatista mountains: In this sense, it is not an overstatement to say that their vision and their word not only resonates with this concept but also has given shape to Dussel's language of transmodernity. The transmodern sense of global modernity opens up the notion that there are many worlds to learn from and relate to beyond a single unitopic imposed order.

Conclusion

> It has a name,
> Creolization, the unstoppable conjunction
> despite misery, oppression, and lynching,
> the conjunction that opens up
> torrents of unpredictable results;
> . . . it is the unpredictability that terrifies
> those who refuse the very idea,
> if not the temptation,
> to mix, flow together, and share.
> —ÉDOUARD GLISSANT, *Faulkner, Mississippi*

The arrogant wish to extinguish a rebellion which they mistakenly believe began in the dawn of 1994. But the rebellion which now has a dark face and an indigenous language was not born today. It spoke before with other languages and in other lands. This rebellion against injustice spoke in many mountains and many histories. It has already spoken in nahuatl, paipai, kiliwa, cucapa, cohcimi, kumiai, yuma, seri, chontal, chinanteco, pame, chichimeca, otomi, mazahua, matlatzinca, ocuilteco, zapoteco, solteco, chatino, papabuco, mixteco, cucateco, triqui, amuzzgo, mazateco, chocho, ixcaateco, huave, tlapaneco, totonaca, tepehua, populuca, mixe, zooque, huasteco, lacandon, mayo, chol, tzeltal, tzotzil, tojolabal, mame, teco, ixil, aguacateco, motocintleco, chicomucelteco. They want to take the land so that our feet have nothing to stand on.... The powerful want our silence. When we were silent, we died, without the word we did not exist.

—ZAPATISTA NATIONAL LIBERATION ARMY, Fourth Declaration
of the Lacandon Jungle

The global battlefield of modernity was structured through processes of ordering and division, emptying and striation, and a spatial logic of concentration around a center. For Hegel, the Mediterranean was the model of the concentration of history: the space where history reached its highest point of dialectical perfection.[48] In contrast, Glissant proposes an archipelagic image of modernity: a poetics of dispersion grounded in the Caribbean, which radiates outward and gestures toward a future not emerging from concentrated historical teleology but from the *irrué*—the sudden, the unforeseeable, the eruptive. The *irrué* also points us to a reading not just of the Caribbean but of the Americas as open and excessive—counter to the condensing, linear logic of the Mediterranean.

Archipelagic thinking offers a politics and poetics of relation from out of an experience of uprooting: a rhizomatic subjectivity that breaks from the plantation, the settlement, the grid town, the colony, or the bounded territory. This mode of subjectivity is not enclosed or abstractly bounded.[49] Instead, the poetics of relation forms intensive connection with one's surroundings, linking up in an archipelagic or scattered way with others. For Glissant, resistance to the destruction of the memory and the roots of African culture, once it crosses over the abyss of the Atlantic in the abyssal belly of the slave boat, is enacted not through a search for an unknown past or a rekindling of tradition. Instead, resistance is enacted through the creativity of relation made possible and grounded in an enmeshed and dispersed network of roots, a network that connects without being subsumed by a center.

Glissant's aesthetic thought offers a model for thinking the role of memory, history, and the projection of the future within uprooted and fragmented space, which has undergone the intensive impulse to be ordered: a response and resistance to the invention of order. The role of forgetting in modernity places the Americas outside the modern matrix, which is essential to the teleological model of modernity in terms of progress and linear concepts supposedly self-generated out of Europe. Glissant, in contrast, offers up a counterhistory or even a nonhistory to show the traumatic destruction of memory and history that was constitutive of the Caribbean: memory not strictly as a practice of something that has to be recovered but something that is to be created through a poetic and creative task.

Indigenous resurgences and resistances stand also in the face of five hundred years of destruction of memory and history. The radical message of having survived the genocidal project of coloniality for five centuries cries forth into the present and the future, not as desire for recognition but as refusal of coloniality. The Zapatista voice from the mountain spirals outward (as a snail's shell, the Zapatista image of the *caracol*), it speaks in many tongues and many lands, "many mountains and many histories." It is a message and rebellion that is not captured by the excessive ordering of coloniality, and one that gathers strength through its horizontal connections to many tongues, struggles, lands, and mountains.

The Americas were constructed both as empty space and as intensively gridded and ordered for purposes of imperial knowledge, control, disciplinary subjection, the excessive violence of fungibility, and extraction of resources. The project of order treated the colonies as a laboratory for the production of new forms of knowledge and new forms of the exertion of power in and by Europe. There is, however, an underside of this history populated with spaces of resistance that formed against or outside the space of order. There is another vision of modernity and an alternative spatiality where subjects link in nonstriated and non-regimented spaces. This is the imaginary of a spatiality that is defined not by its borders and its divisions but by possible modes of relation and connection.

The Caribbean archipelago and the Zapatista mountains embody the excessive landscape of the Americas, terrains of resistance beyond or against the invention of order where another reality grows: a building of worlds in defiance of the impulse toward domination and control. Echoing María Lugones's notion of the hangout carved out of the hierarchized city, or world-traveling in which subjects generate new modes of being, these geographies enact refusal and reinvention.[50] Domination is never complete. From Lugones to Glissant, from Dussel to the Zapatistas, from the sea to the shoal, from the mangrove to the mountain, from the jungle to the city alleyway—there remain ways to redraw the map, to move beyond its imposed lines. The coloniality of emptying and ordering ultimately fails, for something else continues to grow: a resistant modernity, with its poetics, its reason, its modes of relation.

Notes

INTRODUCTION

1. See Portuondo, *Secret Science*.

2. See Portuondo, *Secret Science*, on sixteenth-century Spanish imperial science as "secret science."

3. Borges, "Exactitude," 325.

4. Borges, "Exactitude," 325.

5. Foucault, *Order of Things*.

6. Borges, "Analytical Language," 230. *A* means animal; *ab*, mammalian; *abo*, carnivorous; *aboj*, feline; *aboje*, cat; *abi*, herbivorous; *abiv*, equine; etc.

7. Borges is undoubtedly inspired by the project of a pan-language created by his friend and contemporary Xul Solar.

8. Borges, "Analytical Language," 231.

9. Foucault, *Order of Things*, xv; my emphasis.

10. Foucault does not read Borges's story in terms of the relationship between Europe and the Americas but instead takes up the relationship between the West and the East, and an "exotic" system of thought that is completely inaccessible to European thought. He refers to "our thought" and "our geography," which is broken up by this monstrous classificatory schema of the East.

11. This reading of Borges is one of the only places where Foucault uses the notion of heterotopias in his published writings. The more famous work on heterotopias and questions of spatiality, translated to English as "Of Other Spaces," was originally delivered as a lecture in 1967 and published without review by the author in 1984, shortly before his death. See Foucault, *Order of Things*, xviii. The link between the epistemological space of the heterotopia and the material space of power is one that I explore further in chapter 1. Here I would like to emphasize that this concept emerges in Foucault's thought in relation to Borges, a spatial Latin American thinker.

12. Foucault, *Order of Things*, xix.

13. Borges, "History of Eternity," 131. See also Bosteels, "Borges as Antiphilosopher."

14. For another description of this sense of the New World, see Castro, *Another Face of Empire*, 1–15.

15. See Wynter, "1492." See also Wynter, "How We Mistook the Map for the Territory."

16. See Davenport, "Bull *Inter Caetera*"; and Davenport, "Treaty Between Spain and Portugal."

17. On the notion of abyssal lines, see Santos, "Beyond Abyssal Thinking."

18. See Dussel, *Philosophy of Liberation*, 1–22; and Dussel, *Invention of the Americas*, 12. In the latter text, Dussel describes the "invention" of the Americas as a "covering over" of the other rather than a simple ontological negation.

19. Castro-Gómez offers another reading of Dussel's invention of the other as material practice in "Social Sciences, Epistemic Violence."

20. In his seminal argument in "Coloniality of Power," Aníbal Quijano emphasizes that the emergence of global capitalism (and the coloniality of the modern world-system) was predicated on new practices of labor linked with hierarchical racial classification that were shaped in the colonial encounter with the Americas. On the coloniality of gender and the shaping of gendered spaces in coloniality, see Lugones, "Heterosexualism." For another account of the colonial history of gender, see Marcos, *Taken from the Lips*.

21. Walter Mignolo often claims that developing awareness of the linkage between modernity and coloniality is already a decolonial move. As he writes, "Modernity/coloniality is an imperial package that, of necessity, generates decolonial thinking and action." I take this to mean that one of the first steps of decolonial thought is to highlight the colonial history that is inseparable from modernity. The next step is the "affirmation of the periphery" and the historically excluded—giving voice to subaltern knowledges and modes of existence that have been destroyed or covered over by coloniality. See Mignolo, "Preamble," 17.

22. Santos, "Beyond Abyssal Thinking."

23. On this notion of empty space functioning as the condition of possibility to produce a new spatial infrastructure in the Americas, see also Nemser, *Infrastructures of Race*, 32.

24. Castro-Gómez, *El tonto y los canallas*, 17–42.

25. Wallerstein is himself not a theorist of coloniality but a major influence on Quijano and others with his world-systems theory. See Wallerstein, *Modern World-System I*. Both Dussel and Quijano were major proponents of the Marxist account of global power relations known as dependency theory in Latin America prior to their developments on coloniality. Quijano's major innovation from his earlier work on dependency was to add a robust account of race to account for the colonial nature of labor relations. My argument is not a fundamental departure from Quijano (or Dussel) but more a supplement: to say that we need to look at the local production of power relations and spatial ordering in order to account for global racial classification. Race is produced through local practices and its relay effects with global networks.

26. Castro-Gómez points to the meaning of "hierarchy" as sacralization of power, because it refers to a "sacred authority." His heterarchic account aims to work against sacralizing the power of coloniality by showing that it is not univocal. See Castro-Gómez, "Michel Foucault y la colonialidad," 171. For an English translation, see Castro-Gómez, "Foucault and Coloniality."

27. See Castro-Gómez, *El tonto y los canallas*, 17–48.

28. Lugones, *Pilgrimages/Peregrinajes*, 1–40, 53–64, 77–102, 207–37. See also Certeau, *Practice of Everyday Life*, for an account of strategies and tactics of space (abstracted space versus embodied spaces of lived experience), which Lugones complicates with her notion of tactical strategies.

29. See Rivera, *Andean Aesthetics*, 145–60; and Lugones's account of spatiality and resistance in *Pilgrimages/Peregrinajes*, especially chap. 10, "Tactical-Strategies of the Streetwalker."

30. Kant, *Critique of Pure Reason*, A19/B33–A30/B46.

31. "Captivated by such a proof of the power of reason, the drive for expansion sees no bounds. The light dove, in free flight cutting through the air the resistance of which it feels, could get the idea that it could do even better in airless space." Kant, *Critique of Pure Reason*, A5/B8.

32. In all these thinkers, modernity amounts to a concatenation of events that all take place in Europe. According to this view, modernity is the creation of liberal, individualistic, and industrious subjectivity, along with the capitalist and democratic forms of governance and society, and the triumph of reason over superstition. The crucial events cited by these thinkers are the Protestant Reformation, the scientific revolution, the French Revolution, the Industrial Revolution, and the Enlightenment. Colonialism and the constitution of the first world-system do not play a role in these accounts.

33. See Castro-Gómez, *La hybris del punto cero*, 21–65, for a spatial reading of Kant's "What Is Enlightenment?" and the question of "immaturity." See also Castro-Gómez, *Zero-Point Hubris*.

34. Lefebvre's 1974 book, *The Production of Space*, is a foundational text that puts the organization and production of space at the center of social theory and capitalist relations. His work influences many critical geographers to follow, most notably the Marxist geographer David Harvey in *The Condition of Postmodernity*. See also Elden, *Birth of Territory*, for a genealogical approach to the history of the concept of territory drawing inspiration from the Foucauldian end of the spatial turn.

35. See Mignolo, *Darker Side of the Renaissance*, for an account of the relationship between the colonial shift in cartography and the Renaissance and early Spanish modernity. See Wynter, "1492," on the shift between habitable and uninhabitable space with Columbus's notion of *propter nos* (via *totum navigabile*) that will lead to the emergence of the first modern Rational state-centered image of the human as Man1. See Dussel, *Invention of the Americas*, on the emergence of Europe as global center in response to its conquest of the periphery as the "other" who is covered over.

36. Foucault, "Questions on Geography," 70.

37. In "Foucault and Coloniality," Castro-Gómez argues that we can find an articulation of the relationship between the global, regional, and local levels of power in Foucault's lecture courses from the late 1970s. In this sense, the notion of heterarchy is a concept already in Foucault that Castro-Gómez renders explicit in order to reframe the question of coloniality.

38. O'Gorman, *Invención de América*; Rama, *Lettered City*. There is also a robust tradition of critical cartography and geography within Latin American studies that is important for understanding the coloniality of space. In addition to Portuondo, *Secret Science*,

see Padrón, *Spacious Word*; Padrón, "Mapping Plus Ultra"; Craib, *Cartographic Mexico*; Mundy, *Mapping of New Spain*; Pinet, "Literature and Cartography in Early Modern Spain"; Fraser, *Architecture of Conquest*; Subirats, *El Continente vacío*; Safier, *Measuring the New World*; and Rojas-Mix, *La Plaza mayor*. For the cartographic turn more generally, see Harley, *New Nature of Maps*; Conley, "Early Modern Literature and Cartography"; and Woodward, "Rationalization of Geographic Space."

39. The *long sixteenth century* refers to the second half of the fifteenth century, roughly 1450, to the beginning of the seventeenth century, around 1640. See Wallerstein, *Modern World-System*, vol. 1. This time period captures what Enrique Dussel refers to as the first variant of modernity, prior to the Enlightenment and prior to Descartes. See Dussel, "World-System and 'Trans'-Modernity." Wallerstein's long-term periodization is inspired by Fernand Braudel's *longue durée* conception of spatial histories (geohistories). See Braudel, *La Méditerranée et le monde méditerranéen à l'époque de Philippe II*.

1. ORDERS OF THE GRID

A portion of chapter 1, which has now been expanded and significantly revised, appeared as "Coloniality and Disciplinary Power: On Spatial Techniques of Ordering," in *Inter-American Journal of Philosophy* 10, no. 2 (2019): 25–42.

Epigraph: Sarmiento, *Facundo o Civilización y barbarie*, 23; my translation.

1. Quoted in Kagan, "World Without Walls," 136. See also Tejeira-Davis, "Pedrarías Davila."

2. Defensive architecture did not disappear from all new towns built in the Americas. It would be especially present in the cities of the Caribbean such as San Juan or Havana where defense against corsairs, pirates, and attacks from competing empires made these places especially vulnerable. However, these cases are holdovers and exceptions to the overall process of emergence of a new technique of ordering space.

3. Kagan, "World Without Walls."

4. Foucault refers to these as heterotopias of compensation in one of his few explicit references to colonialism. "Of Other Spaces," 27.

5. Castro-Gómez, "Michel Foucault y la colonialidad."

6. In later chapters, I take up the spatialization of Blackness in early Spanish and Caribbean America through the plantation, the shoal, and the boat.

7. Nemser, *Infrastructures of Race*, 1–23. See also Mills, *Racial Contract*, "Details," 41–90, on the spatialization of race and the racialization of space, especially with respect to the construction of white supremacy as the predominant modern political technique of power.

8. For a review of geographic literature on the grid through the lens of genealogical history, see Rose-Redwood, "Genealogies of the Grid."

9. There are various arguments about the influences and origins of the grid-pattern town and, especially, its extensive deployment in the Americas. Dan Stanislawski, in "Origin and Spread," famously argues that the grid was born in the ancient town of Mohenjo Daro in the ancient Indus civilization and diffused throughout history from there. Others emphasize the Roman Empire and Vitruvius's writings on architecture as

key influences on the Spanish. Others have pointed to the presence of grid patterns in Indigenous American urban planning as a key influence for the Spanish design. Rather than attempting to answer the question of origins, I emphasize the function of this grid pattern in a way that it had not hitherto been deployed.

To be sure, grids are not unique to the sixteenth-century Atlantic. In the sixteenth century, however, the problem of order and the grid is problematized in such an intensive fashion that it gives rise to a new technique for the ordering and epistemology of human space.

10. The notion of the grid that intertwines urban space and writing is developed by Ángel Rama in *Lettered City*. On the coloniality of language, see Veronelli, "Coalitional Approach to Decolonial Communication."

11. Foucault makes a similar methodological operation with respect to the problem of the population as it emerges in the eighteenth century. He points out how the arts of government are blocked between the problem of the state and the problem of the family, but there is no way for governmentality to link the two and be unblocked as its own apparatus until the problem of the population emerges in full swing. *Security, Territory, Population*, 103–4.

12. Foucault, "Of Other Spaces."

13. In his brief consideration of colonialism in relation to heterotopias, Foucault mentions these communities as an example in "Of Other Spaces" (27). For more on the Jesuits in Paraguay, see Ganson, *Guaraní Under Spanish Rule*. This retrospective vision of the Jesuit mission as "utopian" is itself romanticized as Ganson makes clear in her account.

14. See Gómez, *Good Places and Non-Places*.

15. On this latter point, for a reading of the colony as a kind of spatial arrangement that is deeply violent and repressive in its formation but that also gives rise to a new set of spatial practices and resistances, not only repressive but also productive of a new kind of Caribbean subjectivity and modernity, see Glissant, *Poetics of Relation*. In chapter 4 I turn to these questions in Glissant.

16. Rama, *Lettered City*, 1.

17. Aimé Césaire calls this the "boomerang effect" of colonial techniques of power that are later imported back to the metropole (*Discourse on Colonialism*, 36, 41). Foucault echoes Césaire in *Society Must Be Defended* when he describes a boomerang effect that leads to an internal colonialism of Europe. He writes, "A whole series of colonial models was brought back to the West, and the result was that the West could practice something resembling colonization, or an internal colonialism, on itself" (103). This is a moment when Foucault offers a glimpse of how his method of genealogy might trace out colonial modes of power. I pair this moment, which remains underdeveloped in his work, with the suggestive footnote from *Discipline and Punish*, mentioned later.

18. The third domain that he mentions in this note is child-rearing, a topic he lectured on more extensively in *Abnormal* and on which he intended to publish a full volume in original plan for the five volumes of *History of Sexuality*.

19. A number of important works have developed critiques of Foucault and expanded his genealogical methods to account for his omission or limited accounts of the dynamics of race, slavery, and colonialism. For an extensive treatment on Foucault and the questions

of race and colonialism, including a look at how slavery and colonialism create disciplinary spaces especially with respect to sexuality, see Stoler, *Race*. For an account of the flesh and biopolitics with respect to race and colonialism, see Weheliye, *Habeas Viscus*.

20. Foucault, *Discipline*, 314n1.

21. Foucault, *Discipline*, 138.

22. Quijano, "Coloniality of Power."

23. For an excellent reading of Foucauldian accounts of power in relation to the coloniality of power, see Castro-Gómez, "Michel Foucault y la colonialidad."

24. Nemser, *Infrastructures of Race*, 27.

25. Simpson, *Laws of Burgos*, 12; my emphasis.

26. See Lugones, "Heterosexualism."

27. Simpson, *Laws of Burgos*, 22; my emphasis.

28. Viveiros de Castro, *Inconstancy of the Indian Soul*.

29. Vieira, *Sermão do espírito santo* (1657), quoted in Viveiros de Castro, *Inconstancy*, 2.

30. Simpson, *Laws of Burgos*, 18.

31. Simpson, *Laws of Burgos*, 24.

32. Simpson, *Laws of Burgos*, 23.

33. Simpson, *Laws of Burgos*, 19–20.

34. On the encomienda, see Simpson, *Encomienda in New Spain*, 29–38. See also Himmerich y Valencia, *Encomenderos of New Spain*. In the encomienda system, the Amerindians were nominally free subjects but were forced into harsh, unpaid labor in exchange for religious instruction, cultivation, and supposed protection. The system is often linked to medieval feudalism rather than racialized slavery, but it is not clear that it would neatly fit in either model.

35. See Nemser, *Infrastructures of Race*, 25–64, for an extended argument on this question.

36. For a selected translation of these 1573 laws, see Gasparini, "'Laws of the Indies.'" See also Mundigo and Crouch, "City Planning Ordinances." For a complete translation of the ordinances, see Tyler, *Spanish Laws Concerning Discoveries*.

37. Gasparini, "Laws of the Indies," 24.

38. Gasparini, "Laws of the Indies," 24–29.

39. We can see in some of these ordinances how the disciplinary concern of constructing a space and making productive obedient subjects also intersects with the biopolitical concern of producing the life of the population. The care of the Indigenous subject who has been massacred by disease and the violence of conquistadors as well as the livelihood of the Spanish colonists themselves are both at stake in this biopolitical dimension. The Spanish colonial project is not possible as a purely necropolitical one; it must also produce the life of its subjects in the colonies to create a viable political, economic, and spiritual empire. Nemser focuses more on this biopolitical dimension in his *Infrastructures of Race*. For related concerns about biopolitics in sixteenth-century Spanish colonialism via a reading of Bartolomé de las Casas, see Jáuregui and Solodkow, "Biopolitics and the Farming (of) Life."

40. Gasparini, "Laws of the Indies," 27.

41. See Foucault, *Security, Territory, Population*, 20.

42. Foucault, *Security, Territory, Population*, 16.

43. See Rose-Redwood, "Genealogies of the Grid."

44. On intersections of global mapping and state formation, see Biggs, "State on the Map." See also Anderson, *Lineages of the Absolutist State.* On related questions of nationalism, see Anderson, *Imagined Communities*, especially "Creole Pioneers," 47–66, for an account of the vanguard role of independent Latin American nations in the consolidation of the notion of nationalism globally.

45. Foucault, *Security, Territory, Population*, 20.

46. Escobar, "Toward an *urbanismo austríaco*," 171–72.

47. See Foucault, "Of Other Spaces," 27. This is a point at which Foucault provides a very helpful opening to think further the spatial logic of coloniality.

48. It is worth thinking more here about the modes of resistance that are produced and lived out in relation to the imposition of the grid. One example suggested earlier is the forgetfulness that shrugs off this spatiality and its attempted inculcation of virtue. One might then highlight further the internal tensions produced between the heterotopias of order as being constructed by European colonists and Indigenous peoples' understanding and modes of living spatiality that resist this order. Furthermore, it would be worth thinking about what crises and what deviations these heterotopias form in relation to the existing spaces of Europe. I thank one of the anonymous reviewers of an earlier draft of this chapter published in *Inter-American Journal of Philosophy* for their insightful suggestions on heterotopias of deviation and compensation and on the notion of resistance with respect to the existing spaces and modes of living space within Abya Yala (or the Indigenous Americas).

49. These colonial towns and cities also share certain traits with later open cities of circulation and security that will be essential to eighteenth-century industrial capitalism. On circulation in Spanish colonial space, see Nemser, *Infrastructures*, 45–48, 81–86. Colonial circulation prefigures many of these later technologies with emphases on control through the openness and distribution of gridded space rather than the enclosure of the wall or the fortress. The colonial gridded town or city can also be considered as one node in a larger network of colonial circulation. Foucault describes this problem of circulation and city walls as follows: "And finally, an important problem for towns in the eighteenth century was allowing for surveillance, since the suppression of city walls made necessary by economic development meant that one could no longer close towns in the evening or closely supervise daily comings and goings, so that the insecurity of the towns was increased by the influx of the floating population of beggars, vagrants, delinquents, criminals, thieves, murderers, and so on, who might come, as everyone knows, from the country. In other words, it was a matter of *organizing circulation*." *Security, Territory, Population*, 18.

50. Castro-Gómez, *Zero-Point Hubris*, 218.

51. Sarmiento, *Facundo o Civilización y barbarie*, 23; my emphasis.

52. In chapter 2, I develop on this question of smooth and striated space. See also Chaunu, *L'expansion européenne*; and Deleuze and Guattari, "1440," in *A Thousand Plateaus*, 474–500.

53. For a genealogy of the issues of moral and physical geography in the preceding period of 1750–1816 in the criollo discourses of New Granada in relation to Enlightenment European discourses, see Castró-Gomez, *La hybris del punto cero*, 228–303.

54. Sarmiento, *Facundo o Civilización y barbarie*, 28; my translation.

55. See Sarmiento, *Conflicto y armonías*. See also Hooker, *Theorizing Race in the Americas*.

56. See Wynter, "1492."

2. ORDERS OF MOVEMENT

1. See Kotef, *Movement and the Ordering of Freedom*; and Kotef, "Movement." As Kotef points out, the chief example of this equation of movement and freedom in the liberal tradition is Hobbes's *Leviathan*, in which "life is but a motion of Limbs," and we seek liberty as unconstrained movement (9). The right to liberty is essentially to move as we please.

The main theorist of the bounding of the land-property relation is Locke in the *Second Treatise*, §§ 25–51. See also Rousseau's critique of property, in which the origin of evil stems from the first person who encloses a plot of land and claims, "This is mine"; "Discourse on the Origins," 161. Vitoria's "De Indis," or "On the American Indians," can be read as a precursor to these problematizations of freedom, property, and movement (in relation to the common earth, which as Rousseau claims, "belongs to us all"); see Vitoria, *Political Writings*.

2. Primitive accumulation is precisely the articulation between movement and settlement-extraction that is predicated on the coloniality of space. On Indigenous dispossession and primitive accumulation (theft) as the generative mechanism of property relations, see Nichols, *Theft Is Property!* On the theme of Indigenous dispossession and primitive accumulation, see also Coulthard, *Red Skin, White Masks*.

3. Byrd, *Transit of Empire*.

4. Lugones, *Pilgrimages/Peregrinajes*, 8–16.

5. Later racial geographies (as evidenced in Sarmiento, Kant, Comte de Buffon, Cornelius de Pauw, and others) will expand on this idea by developing geographic determinism, in which a certain land and climate produces morally and racially deficient subjects. When the determinism is not absolute, the ability to escape it is also often racialized. I refer here to the eighteenth- and nineteenth-century developments of geography as a scientific discipline and its connections to scientific racism. On this question, see Castro-Gómez, *Zero-Point Hubris*, chap. 5, 197–265.

6. See Mair, *In Secundum Sententiarum*.

7. See Sepúlveda, "Democrates Part Two."

8. On this notion of the spatialization of race, see Wynter, "Unsettling"; McKittrick, *Demonic Grounds*; Mills, *Racial Contract*; and Nemser, *Infrastructures of Race*.

9. See Wynter, "Unsettling." See also Schmitt, *Concept of the Political*.

10. Wynter, "Unsettling," 264.

11. On this question, see also Wynter, "Pope Must Have Been Drunk."

12. Santos, "Beyond Abyssal Thinking."

13. On global lines, or *rayas*, see Schmitt, *Nomos of the Earth*.

14. See Schmitt, *Nomos of the Earth*.

15. McKittrick, *Demonic Grounds*, 129. Wynter uses the notion of Man1 to refer to the first period of the racialization of the human after the conquest of the Americas and

commencement of the transatlantic African slave trade in the fifteenth and sixteenth centuries. Man1 is the overrepresentation of the European white man as human in the triadic model of European, African, and (Native) American. This is also the emergence of a Rational conquering Man of the State who can navigate a world made for them (*propter nos*) against a fallen flesh trapped in a corrupt materiality. Man1 is prior to the biological and quasi-scientific racism of the nineteenth century, which she will refer to as Man2. See Wynter, "Unsettling," 264.

16. Hanke, *All Mankind Is One*.

17. On reading Las Casas as an imperialistic thinker, see Castro, *Another Face of Empire*; and on Las Casas and race, see Von Vacano, "Paradox of Empire."

18. The relationship and influence between Vitoria and Las Casas are significant and complex. They write from different positions of training, experience, and authority but their arguments also often echo each other. They are both Dominicans (a considerable overlap in their respective formations), but Vitoria writes from within the university in Spain, while Las Casas writes from his experience in the Americas. It is important to note that Las Casas's work goes beyond Vitoria's approach to Indigenous self-determination in his later writing, where he argues in favor of Peruvian sovereignty. See Las Casas, *Doce Dudas*; see also Zorrilla, "Just War"; and Brunstetter, "Las Casas."

19. Vitoria, *Political Writings*, 253.

20. For an excellent critique of Vitoria's supposedly humanist legacy as a masked imperialism, see Anghie, *Imperialism*. See also Schmitt, *Nomos of the Earth*, for a key reading of Vitoria's role in shaping the new nomos of the earth, post 1492. See also Bohrer, "Color-Blind Racism in Early Modernity."

21. Vitoria, *Political Writings*, 151. See also Vera, "Papal Bull to Racial Rule," 457.

22. Vitoria, *Political Writings*, 278.

23. Vera, "Papal Bull to Racial Rule," 454.

24. Schmitt, *Nomos of the Earth*, 86. See also Scott, *The Catholic Conception of International Law*.

25. Sloterdijk, *In the World Interior*, 3–14. See also Sloterdijk, *Globes*.

26. Sloterdijk, *In the World Interior*, 30.

27. On the "apparatus of capture," see Deleuze and Guattari, *A Thousand Plateaus*, 424–73; on the complex relation between smooth and striated space, 474–75.

28. Schmitt, *Nomos of the Earth*, 42.

29. Schmitt, *Nomos of the Earth*, 42.

30. Schmitt, *Nomos of the Earth*, 43.

31. Deleuze and Guattari, *A Thousand Plateaus*, 380.

32. Deleuze and Guattari, *A Thousand Plateaus*, 380.

33. For another critical reading of the modern nomos inspired by Schmitt, see Galli, *Political Spaces and Global War*. Agamben's *Homo Sacer* must also be considered in this dialogue for his account of the concentration camp as "the *nomos* of the modern." See also Heller-Roazen, *Enemy of All*, for an account of piracy in relation to the modern nomos. Weheliye's *Habeas Viscus* takes up a critical reading of Agamben's claim via the question of racialization.

34. Deleuze and Guattari, *A Thousand Plateaus*, 351.

35. Schmitt, *Nomos of the Earth*, 94.

36. Deleuze and Guattari, *A Thousand Plateaus*, 385.

37. Vitoria, *Political Writings*, 278.

38. Vitoria, *Political Writings*, 278. In a similar way, Kant will argue that the right to hospitality cannot be used as a justification for colonial conquest or land appropriation—yet Vitoria performs exactly this conceptual slippage. See Kant, "Perpetual Peace."

39. Quoted in Pagden, *Spanish Imperialism*, 24.

40. Lugones, *Pilgrimages/Peregrinajes*, 9.

41. Like Cano and in a more radical fashion, Bartolomé de Las Casas will engage in several such thought experiments that reverse the colonized-colonizer directionality to envision how Spain might react if the Natives from America made just war claims against them.

42. See Coulthard, *Red Skin, White Masks*.

43. Glissant, *Poetics of Relation*.

44. Davis, *Choose Your Bearing*.

45. Glissant, *Poetics of Relation*, 14.

46. Hubbard, *Happiness of a People*.

47. It should be noted that this was first a legal and political argument made by the English government before it was appropriated by religious groups. The Crown had claimed *vacuum domicilium* of the land between 40° and 48° latitude, claiming that the land was totally uninhabited even though there were still some Natives in that region. See Neuwirth, "Images of Place," 44. Undoubtedly, the Puritans had to make this argument their own and fortify it for their appropriation of the land to be religiously palatable.

48. Cotton, "God's Promise," 4.

49. This might also be because much of the land was not actually vacant, only unenclosed. This use of enclosure as the only valid form of land ownership and appropriation will be crucial for both the Puritans and later for Locke.

50. Cotton, "God's Promise," 5; my emphasis.

51. Carroll, *Puritanism and the Wilderness*, 17–18.

52. Neuwirth, "Images of Place," 42–53.

53. These categories are drawn from Neuwirth, "Images of Place," 43.

54. Neuwirth, "Images of Place," 48–51.

55. Locke, "Of Property," in *Second Treatise*, §§ 42, 26.

56. Locke, "Of Property," §§ 37, 24.

57. I draw this term from Rivera's discussion of colonial and Andean spatiality in *Andean Aesthetics*, 145–47.

58. Lugones, *Pilgrimages/Peregrinajes*, 8.

59. Lugones, *Pilgrimages/Peregrinajes*, 9.

60. Lugones, *Pilgrimages/Peregrinajes*, 11.

61. Lugones, *Pilgrimages/Peregrinajes*, 76.

62. See Rivera, *Andean Aesthetics*, 146–47.

63. Glissant, *Poetics of Relation*, 12.

64. Lugones, *Pilgrimages/Peregrinajes*, 11.

65. See Gualdrón Ramirez, "To 'Stay Where You Are,'" for a reading of Glissant in terms of a politics of place as a kind of affirmation without return or detour. Perhaps

this might also constitute a kind of trespassing against dominant sense by just "stay[ing] where you are."

66. Simpson, *Mohawk Interruptus*, 11.

67. On this usage of *settled*, see Simpson, *Mohawk Interruptus*, 11.

3. TRANSMODERNITY AND THE BATTLEFIELD OF COLONIALITY

A portion of chapter 3, expanded and significantly revised, will appear as "Transmodern Geographies and Coloniality: On Enrique Dussel's Pluriversal Modernity," in *Latin American Perspectives*, March 2026, in their special issue on Aníbal Quijano.

Epigraphs: This biblical passage (Luke 14:23; my emphasis) is cited by Saint Augustine to justify Donatist coercion and is mobilized to similar ends here by Sepúlveda. The same passage is cited by Foucault in *History of Madness* (42) at the outset of his analysis of the great confinement of the mad and the poor across Europe in the seventeenth century. Foucault, *History of Madness*, xxviii. Theodor Adorno, *Negative Dialectics*, 302. Édouard Glissant, *Philosophie de la Relation*, 62–63; my translation.

1. Dussel, *Philosophy of Liberation*, 1–15.

2. The most well-known exemplar of this approach is Galeano's bombshell 1971 text, *Open Veins of Latin America*. This work builds on a larger tradition that began to be carved out as early as the 1950s in the structuralist work of the Argentine economist Raúl Prebisch.

3. For an insightful reading of these two concepts in Dussel, see Vallega, *Latin American Philosophy*, 68–74.

4. See Vallega, *Latin American Philosophy*, 68–74, for a development of this notion as a form of aesthetics and preconceptual sensibility.

5. This notion is also essential to decolonial aesthetics as I argue in Deere, "Spacing of Decolonial Aesthetics."

6. This periodization of two stages of modernity argues that there is a colonial stage of modernity prior to the standard claims of a mid-seventeenth-century origin with Descartes and the scientific revolution in Europe. Instead, for Dussel there is a first colonial stage of modernity wherein this process of European universalization begins and the constitution of the first world-system comes into place. While I believe this is a helpful corrective to the standard Eurocentric account and periodization of modernity, I think it runs the risk of reproducing some of the same problems it aims to critique: namely, dividing up two careful periods and separating the first colonial stage from a second stage that is more exclusively European.

7. As I advance this decolonial metanarrative, it is also important to issue a reminder that I later return to local practices of power and knowledge. Furthermore, knowledge does not simply issue from power in the process of conquest, rather they follow parallel pathways that intersect, interrupt, increase, and influence one another. Power will transform knowledge, and knowledge will transform power in this history. I follow Santiago Castro-Gómez in the view that one can advance a *heterarchic* conception of power that tracks the local, global, and intermediate ranges of power relations.

8. This periodization comes from the French historian Fernand Braudel, who coined the notion of the *longue durée*. His innovations in the Annales school would

conceptualize large-scale historical periods around geographic spatial analysis. He is another major influence, in addition to dependency theory, for Immanuel Wallerstein's world-systems analysis. See Lee, *Longue Durée*.

9. Dussel, *Philosophy of Liberation*, 1.

10. This list would also later include the Jesuit philosopher Francisco Suárez, probably the greatest philosopher and theologian of his era up until Descartes.

11. Dussel, *Philosophy of Liberation*, 2.

12. The role of presences and absences in phenomenology certainly comes to mind here. Yet every absence is always a potential present in this view, while the other, or the space of nonbeing, is not considered in a potential present but rather the radical outside.

13. Dussel, *Philosophy of Liberation*, 6.

14. I think Dussel could actually be read here not as rejecting phenomenology wholesale but rather as a sort of critical phenomenologist. That is to say, the very grounds of lived experience should not be taken for granted as natural or pregiven. Rather they are dimensions that have been shaped historically by knowledge and power.

15. Dussel, *Philosophy of Liberation*, 2. If we read this passage together with the heterarchical local reading of spaces, we could also account for the local distributions of this problem within a place like New York City itself: The differences between being born near Central Park in Manhattan versus the Southeast Bronx would also be quite immense.

16. However, I would avoid claiming that ontology is just reified epistemology. It is not simply a one-way street. Again, they intersect, and ontological transformations of spaces and subject also shift the geopolitics of knowledge.

17. Descartes, *Meditations on First Philosophy*.

18. My interest here is not a strict exegesis of the internal arguments of his text but rather an extraction of the basic characteristics of this Cartesian figure of thought. It is this Cartesian figure, the Cartesian subject that lives beyond the arguments of Descartes's texts, that has had the most persistent effect on Western ontology and epistemology.

19. Beauvoir's critique of Sartre is precisely predicated on a richer thinking of the subject's relation to others and the subject's entanglement with and responsibility for the freedom of others. This is precisely a thinking that aims to avoid the threat of solipsism. Beauvoir, thus, would actually be a richer interlocutor with Dussel than Sartre, as they both offer a sort of critical phenomenology (Beauvoir, *Ethics of Ambiguity*). Beauvoir writes in the conclusion that her ethics is individualistic in some sense as the individual is the source of freedom and site of responsibility; however, "it is not solipsistic, since the individual is defined only by his relationship to the world and to other individuals; he exists only by transcending himself, and his freedom can be achieved only through the freedom of others" (156).

20. Descartes, *Meditations on First Philosophy*, 16.

21. Castro-Gómez, *Zero-Point Hubris*.

22. See Castro-Gómez, *Zero-Point Hubris*, chap. 1, for a very interesting development of this point that also develops on Dussel's insights into the dominating nature of the *ego cogito*.

23. Dussel, *Philosophy of Liberation*, 3.

24. Two such examples are Castro-Goméz, *Crítica de la razón latinoamericana*; and Maldonado-Torres, *Against War*.

25. This critique has been developed most forcefully in one of the first English-language book-length works on Latin American philosophy: Schutte, *Cultural Identity and Social Liberation*.

26. Cerruti, *Filosofía de la liberación*.

27. This early critique of liberation philosophy has been returned to in various ways more recently by thinkers such as Schutte, Castro-Gómez, and Maldonado-Torres.

28. Already it is clear how dependency theory questions the developmentalist narrative and the temporalization of Latin American space as primitive. This is a key innovation of Latin American thought that opens up the question of considering the spatial nature of domination. Immanuel Wallerstein is also crucially influenced by this theoretical innovation when he develops world-systems analysis. For a further analysis of this question—namely, dependency and developmentalist—see Grosfoguel, "Developmentalism, Modernity, and Dependency."

29. Castro-Gómez summarizes this critique at the outset of the first chapter of *Crítica de la razón latinoamericana*. The English translation of this first chapter is "Challenge of Postmodernity," in *Critique of Latin American Reason*.

30. Though positivism is usually attributed to Auguste Comte, it must not be forgotten that the most fervent practice and development of positive doctrine took place in Latin America, especially in Brazil, Argentina, and Mexico. The Brazilian flag still bears the positivist motto "Order and Progress."

31. Schutte, *Cultural Identity and Social Liberation*.

32. Maldonado-Torres, *Against War*, 163–65.

33. This concern is raised also by Castro-Gómez, *Crítica de la razón latinoamericana*.

34. Dussel, *Ethics of Liberation*.

35. See Alcoff, "Philosophy, Conquest, and the Meaning of Modernity," and Alcoff, "Enrique Dussel's Transmodernism," for excellent accounts of Dussel's transmodernism that clearly navigate around this trap of absolutism.

36. For an excellent development of this notion, see Silva, "Americas Seek Not Enlightenment."

37. As a related remark, I suggest that it would be a mistake to read Dussel as simply writing commentaries on other European thinkers such as Levinas, Heidegger, or Marx. Instead, one must be attentive to the creation of new modes of thought that emerge from such a confrontation of sources alongside the struggles that he is engaged with in Latin America.

38. I elaborate on this theme later in considering the dyad between center and periphery: This is a dyad that must be left behind in the final analysis as it reproduces the binary distinctions that are endemic to modern thought, and it polarizes the two forms of subjectivity against one another. Here, again, I think Dussel offers resources that take us in this direction, but he does not always go far enough.

39. See Dussel, "Agenda."

40. This last phrase comes from the Zapatista liberation struggle in Chiapas, the southeast region of Mexico: "Un mundo donde quepan muchos mundos."

41. Dussel, *El encubrimiento del otro*, 8; my translation.

42. In the original Spanish, the title is *El encubrimiento del otro*. The notion of eclipse used by the translator, Michael Barber, captures nicely the image of misrecognition at work in the term *en-cubrimiento*, but it does not capture the wordplay with the notion of discovery (*des-cubrimiento*) that was traditionally used to describe the 1492 event.

43. It is worth highlighting that Dussel is speaking in Frankfurt as he delivers these lectures: Many themes in the lectures echo those of the first Frankfurt school but are taken from a decolonial perspective that thinks the history of colonization as central to the history of instrumental reason, on the five-hundred-year anniversary of 1492.

44. I think this was also the case in my analysis of the critiques of Dussel in this chapter's section "A Material Account of the Other."

45. See Castró-Gomez, "Social Sciences, Epistemic Violence."

46. In Ángel Rama's work along with the thought of Domingo F. Sarmiento, we can see how a center was built within the periphery, what Rama calls the "lettered city." Rama, *Lettered City*; and Sarmiento, *Facundo*. This internal colonization would impose a center-periphery binary within Latin American nations themselves, between the supposed civilization of the cities and the barbarism of the countryside. The peripheral zones produced within spaces of the center, however, must also not be forgotten, though I will not have the space to address these in this book. Latinx and Chicanx struggles in Los Angeles, African American struggles for liberation, North African migrants in the suburbs of Paris, or even Antonio Gramsci's example of the extreme poverty and marginalization of the south of Italy with respect to the north are all examples of peripheries within the center (or the south in the north). See Gramsci, "Some Aspects of the Southern Question."

47. Dussel, *Ethics of Liberation*, 25.

48. Dussel, *Ethics of Liberation*, 25.

49. Dussel, *Ethics of Liberation*, 55–107, 215–90. He explicitly takes up Castro-Gómez's critique and emphasizes this materiality in a footnote (470n222).

50. Dussel, *Philosophy of Liberation*, viii, first published in 1977. Lyotard's *Postmodern Condition* would not be published in French until 1979.

51. See Vallega, *Latin American Philosophy*, 81–95, for a critique of Dussel on similar grounds of setting the terms of reason in the language of the center, one that was illuminating for my own thinking on this question. As a corrective to this subordination, Vallega points to the need for an aesthetics of liberation to account for the creative praxis of the excluded other. This point is well taken, as Dussel does, indeed, turn to an aesthetics of liberation in his late and final works.

52. On the coloniality of language, see Veronelli, "Coalitional Approach to Decolonial Communication." On the question of listening, decoloniality, and histories of violence, see Acosta López, "Gramáticas de lo inaudito as Decolonial Grammars." For the first Spanish-language grammar and the first grammar of any Latinate vernacular language, see Nebrija's 1492 gift to Queen Isabella, *Gramática de la lengua castellana*.

53. Dussel, *Posmodernidad y transmodernidad*. See also Vattimo, *End of Modernity*.

54. Dussel, *Invention of the Americas*, 26.

55. I am accepting the large label of *postmodern* here to refer to a number of different thinkers who do not necessarily identify as such. There are nuances that should be introduced, but for the sake of referring to a number of Continental thinkers who are

generally critical of the modern project in one way or another, I tentatively employ the term. However, the argument I am making here should also show that there are complex differences among a number of thinkers who may or may not embrace this term. Here, he includes thinkers ranging from Lyotard, Vattimo, Derrida, Rorty, Foucault, to Adorno. On Rorty and others, see also Dussel, *The Underside of Modernity*. See Derrida, *Writing and Difference*; and Rorty, *Philosophy and the Mirror of Nature*.

56. It could be added that the blanket rejection of all postmodern thought as irrationalist too quickly dismisses the nuances and resources of these modes of critique. Further, the nihilistic emphasis in Vattimo might be distinguished from the critical project found in Foucault or Adorno.

57. See Vallega, *Latin American Philosophy*, 81–95.

58. For a more general account of the possible dialogues and divergences between the work of Foucault and Dussel, see Alcoff, "Power/Knowledges."

59. Foucault, *History of Madness*, xxviii.

60. Huffer discusses this passage in detail in relation to questions of queer theory in *Mad for Foucault*.

61. Dussel, *Invention of the Americas*, 9. The influence of Fanon on Dussel's philosophy of liberation is notable here. See Fanon, *Wretched of the Earth*, 113.

62. In his late work, however, Foucault refuses the blackmail of being for or against the Enlightenment and claims it is part of the historical ontology of the present with which one has to work and critique. Thus, it would also be misleading to situate Foucault's work as antimodern or anti-Enlightenment. See Foucault, "What Is Enlightenment?"

63. Dussel also aims for transmodernity and the philosophy of liberation to enter into an intercultural (North-South and South-South) dialogue that has been foreclosed by modernity. His critique of late Frankfurt school thinkers, such as Karl-Otto Apel and Jürgen Habermas along with Anglophone political theorists such as John Rawls, is that they do not recognize the fundamental asymmetry that is in place and that would render any true dialogue or social contract possible. Dussel hopes to expose the roots of this asymmetry and domination and also uncover the history of the other that has been covered over: "The philosophy of liberation begins by affirming alterity, but it also recognizes negative aspects such as the concrete, empirical impossibility of the excluded or dominated one ever being able to intervene *effectively* in dialogue." *Invention*, 12–13.

64. Castro-Gómez, "Social Sciences, Epistemic Violence," 270.

65. Maldonado-Torres, "Topology of Being," 29.

66. Castro-Gómez, *Crítica de la razón latinoamericana*, 166.

4. ARCHIPELAGOS OF RESISTANCE

Epigraphs: Benítez-Rojo, *Repeating Island*, 2; translation modified. Glissant, *Caribbean Discourse*, 11. Benítez-Rojo, *Repeating Island*, 21. Glissant, *Faulkner, Mississippi*, 11. Glissant, *Poetics of Relation*, 187. Bernabé, Chamoiseau, and Confiant, *Éloge de la Créolité*, 89.

1. Glissant, *Introduction to a Poetics of Diversity*, 3–4.

2. The tripartite notion of the abyss is further developed in Glissant, *Poetics of Relation*, 5–9.

3. For an illuminating account of Glissant's creolization in dialogue with Lugones and Latina feminists such as Gloria Anzaldúa and Mariana Ortega, see Sealey, *Creolizing the Nation*. See also Ortega, *In-Between*; and Pitts, *Nos/Otras*, for readings of Anzaldúa and Lugones that engage with questions of multiplicity that echo some aspects of Glissant's notion of creolization.

4. Simpson, *As We Have Always Done*, 9.

5. Kris Sealey's work points to this notion of cacophony in Byrd's work in dialogues with Glissant's creolization. See her excellent article on the entanglements and incommensurability of Black and Indigenous resistance read through these notions, "When Heads Bang Together."

6. Byrd, *Transit of Empire*, xxvii.

7. Byrd, *Transit of Empire*, 20. The notion of *haksuba* also echoes the Nahua concept of *nepantla* as balancing the tension between two extremes when living in a world in motion. Gloria Anzaldúa develops on *nepantla* and its "cracks between worlds" that can generate resistance in very fruitful ways in *Light in the Dark*, 71–74. On the importance of *nepantla* in Aztec/Nahua philosophy, see Maffie, *Aztec Philosophy*, 523–29.

8. Byrd's principle of cacophony is akin to Santiago Castro-Gómez's notion of coloniality as a heterarchy of power, shifting away from a top-down, binary account of colonialism. As I discuss with this notion of heterarchy in chapter 1, this shift away from a top-down view of coloniality also enables a richer account of the cracks of resistance and the local grounds from which resistances emerge.

9. For a clear account of both this incommensurability and entanglement of resistance, see Sealey, "When Heads Bang Together."

10. Sealey, "When Heads Bang Together."

11. King, *Black Shoals*.

12. For a key account of Blackness and liquidity, see Sharpe, *In the Wake*.

13. For an excellent account of the role of this abyssal beginning at the core of Glissant's philosophy, see Drabinski, *Glissant and the Middle Passage*.

14. In *What Is Philosophy?*, Deleuze and Guattari develop the term *geophilosophy* in a way that contains rich significations for the spatialization of thought that I trace out in this and previous chapters. However, they remain tied to a Hellenocentric historiography (and by extension, Eurocentric), tying the origins of philosophy to the geographic flows and intensities that surrounded and populated ancient Athens and ancient Greece. It is worth remarking, however, that what we call ancient Greece is also an archipelago.

15. For a history of the Plantation in the Caribbean, which links and differentiates its histories in the islands colonized by the Spanish, French, Dutch, and English, see Benítez-Rojo, *Repeating Island*, 33–81.

16. In the sixteenth century both institutions coexisted, while the Plantation did not take on its full-fledged form until the seventeenth and eighteenth centuries as French, English, and Dutch colonists began to take over Caribbean islands that Spain had mostly abandoned after the sugar recession of the early seventeenth century. See Benítez-Rojo, *Repeating Island*, 38–48.

17. See Glissant, *Caribbean Discourse*, 162–70; see also Glissant, *Introduction to a Poetics of Diversity*, 19–36.

18. See Glissant, *Poetics of Relation*, epigraph; and Glissant, *Treatise on the Whole-World*, 45–54.

19. Bernabé, Chamoiseau, and Confiant, "In Praise of Creoleness," 901. Glissant makes important departures from the concept of *creolité* offered by Bernabé, Chamoiseau, and Confiant. Primarily, Glissant prefers a nonsubstantive notion of *creolization* rather than *creoleness*, to capture the ever-unfolding and nonfixed character of this movement. My aim here is not to unpack all these debates but rather to unpack some of their resonances for thinking through the question of the coloniality of space.

20. Glissant, *Poetics of Relation*, 73–75.

21. Glissant, *Poetics of Relation*, 74.

22. On the rhizome, see also Deleuze and Guattari, *A Thousand Plateaus*, 3–25. For a reading of the Caribbean mangrove as Glissantian rhizome, see Drabinski, "Poetics of the Mangrove."

23. Deleuze and Guattari, *A Thousand Plateaus*, 25.

24. Rhizomatic relation allows also an avenue to extend Dussel's notion of philosophy from the underside of modernity in terms of a geophilosophical concept of the nonhierarchical and noncentered.

25. Glissant, *Poetics of Relation*, 11.

26. The connection between colonialism and state formation is also one that is developed in chapter 2 through the notion of the nomos of the earth and the bounding of smooth space.

27. Glissant, *Poetics of Relation*, 15.

28. Glissant, *Poetics of Relation*, 14.

29. Glissant, *Poetics of Relation*, 12

30. Glissant's analysis here connects in interesting ways with the discussion of the balance between smooth and striated space (or free and ordered space) as constitutive of the formations of the modern state form discussed in chapter 2 of this book, in particular.

31. Glissant, *Poetics of Relation*, 18.

32. Glissant, *Poetics of Relation*, 18.

33. Glissant, *Poetics of Relation*, 18.

34. Glissant, *Poetics of Relation*, 18.

35. Glissant, *Poetics of Relation*, 144.

36. See Glissant, *Caribbean Discourse*, 62.

37. Glissant, *Faulkner, Mississippi*, 1.

38. Glissant, *Poetics of Relation*, 8.

39. Glissant, *Poetics of Relation*, 75.

40. Glissant, *Poetics of Relation*, 64.

41. Glissant, *Poetics of Relation*, 73.

42. Glissant, *Poetics of Relation*, 1.

43. Glissant, *Poetics of Relation*, 189.

44. Glissant, *Poetics of Relation*, 144.

45. Dussel, *Ethics of Liberation*.

46. Indigenous Revolutionary Clandestine Committee of the Zapatista National Liberation Army, "Tomorrow Begins Today."

47. Marcos, "Fourth Declaration."

48. Hegel, *Lectures on the Philosophy of History.*

49. For another sense of resistance to the coloniality of space within its tendencies to bound and contain, see Zambrana's reading of the shipping container in Puerto Rico through the lens of decoloniality as reparations in *Colonial Debts.*

50. Lugones, *Pilgrimages/Peregrinajes.*

Bibliography

Acosta López, María del Rosario. "Gramáticas de lo inaudito as Decolonial Grammars: Notes for a Decolonization of Listening." *Research in Phenomenology* 52, no. 2 (2022): 203–22.

Adorno, Theodor. *Negative Dialectics*. Translated by E. B. Ashton. New York: Continuum, 1978.

Agamben, Giorgio. *Homo Sacer: Sovereign Power and Bare Life*. Translated by Daniel Heller-Roazen. Stanford, CA: Stanford University Press, 1998.

Alcoff, Linda Martín. "Enrique Dussel's Transmodernism." *Transmodernity: Journal of Peripheral Cultural Production of the Luso-Hispanic World* 1, no. 3 (2012): 60–68.

Alcoff, Linda Martín. "Philosophy, the Conquest, and the Meaning of Modernity: A Commentary on 'Anti-Cartesian Meditations: On the Origin of the Philosophical Anti-Discourse of Modernity' by Enrique Dussel." *Human Architecture: Journal of the Sociology of Self-Knowledge* 11, no. 1 (2013): 57–66.

Alcoff, Linda Martín. "Power/Knowledges in the Colonial Unconscious: A Dialogue Between Dussel and Foucault." In *Thinking from the Underside of History: Enrique Dussel's Philosophy of Liberation*, edited by Linda Martín Alcoff and Eduardo Mendieta, 249–68. Lanham, MD: Rowman and Littlefield, 2000.

Anderson, Benedict. *Imagined Communities: Reflections on the Origin and Spread of Nationalism*. New York: Verso, 2006.

Anderson, Perry. *Lineages of the Absolutist State*. New York: Verso, 2013.

Anghie, Anthony. *Imperialism, Sovereignty, and the Making of International Law*. Cambridge: Cambridge University Press, 2005.

Anzaldúa, Gloria. *Light in the Dark/Luz en lo oscuro*. Edited by Ana Louise Keating. Durham, NC: Duke University Press, 2015.

Beauvoir, Simone de. *The Ethics of Ambiguity*. New York: Citadel Press, 1948.

Benítez-Rojo, Antonio. *La isla que se repite: El Caribe y la perspectiva posmoderna*. Hanover, NH: Ediciones del Norte, 1989.

Benítez-Rojo, Antonio. *The Repeating Island: The Caribbean and the Postmodern Perspective*. Translated by James E. Maraniss. Durham, NC: Duke University Press, 1996.

Bernabé, Jean, Patrick Chamoiseau, and Raphaël Confiant. *Éloge de la Créolité/In Praise of Creoleness*. Translated by M. B. Taleb-Khyar. Paris: Éditions Gallimard, 1989.

Bernabé, Jean, Patrick Chamoiseau, and Raphaël Confiant. "In Praise of Creoleness." *Callaloo* 13, no. 4 (1990): 886–909.

Biggs, Michael. "Putting the State on the Map: Cartography, Territory, and European State Formation." *Society for Comparative Study of Society and History* 41, no. 2 (1999): 374–405.

Bohrer, Ashley J. "Color-Blind Racism in Early Modernity: Race, Colonization, and Capitalism in the Work of Francisco de Vitoria." *Journal of Speculative Philosophy* 32, no. 3 (2018): 388–99.

Borges, Jorge Luis. "The Analytical Language of John Wilkins." In *Selected Non-Fictions*, edited by Eliot Weinberger, translated by Esther Allen, Suzanne Jill Levine, and Eliot Weinberger, 229–32. New York: Viking, 1999.

Borges, Jorge Luis. "The History of Eternity." In *Selected Non-Fictions*, edited by Eliot Weinberger, translated by Esther Allen, Suzanne Jill Levine, and Eliot Weinberger, 111–18. New York: Viking, 1999.

Borges, Jorge Luis. "On Exactitude [Del Rigor] in Science." In *Collected Fictions*, translated by Andrew Hurley, 325. New York: Penguin, 1998.

Bosteels, Bruno. "Borges as Antiphilosopher." *Vanderbilt E-Journal of Luso-Hispanic Studies* 3 (2006). https://doi.org/10.15695/vejlhs.v3i0.3197.

Braudel, Fernand. *La Méditerranée et le monde méditerranéen à l'epoque de Philippe II.* Vol. 1. Paris: Armand Colin, 2017.

Brunstetter, Daniel. "Las Casas and the Concept of Just War." In *Bartolomé de las Casas, O.P.: History, Philosophy, and Theology in the Age of European Expansion*, edited by David Thomas Orique, OP, and Rady Roldán-Figueroa, 218–42. Leiden, Netherlands: Brill, 2019.

Byrd, Jodi. *The Transit of Empire: Indigenous Critiques of Colonialism.* Durham, NC: Duke University Press, 2011.

Carroll, Peter N. *Puritanism and the Wilderness.* New York: Columbia University Press, 1969.

Castro, Daniel. *Another Face of Empire: Bartolomé de las Casas, Indigenous Rights, and Ecclesiastical Imperialism.* Durham, NC: Duke University Press, 2007.

Castro-Gómez, Santiago. *Crítica de la razón latinoamericana.* Barcelona: Puvill Libros, 1996.

Castro-Gómez, Santiago. *Critique of Latin American Reason.* Translated by Andrew Ascherl. New York: Columbia University Press, 2021.

Castro-Gómez, Santiago. *El tonto y los canallas: Notas para un republicanismo transmoderno.* Bogotá: Editorial Pontificia Universidad Javeriana, 2019.

Castro-Gómez, Santiago. *La hybris del punto cero.* Bogotá: Universidad Javeriana, 2010.

Castro-Gómez, Santiago. "Michel Foucault and the Coloniality of Power." Translated by Kyle Kopsick and David Golding. *Cultural Studies* 37, no. 3 (2023): 444–60.

Castro-Gómez, Santiago. "Michel Foucault y la colonialidad del poder." *Tabula Rasa* 6 (2007): 153–72.

Castro-Gómez, Santiago. "The Social Sciences, Epistemic Violence, and the Problem of the 'Invention of the Other.'" *Nepantla: Views from the South* 3, no. 2 (2002): 269–85.

Castro-Gómez, Santiago. *Zero-Point Hubris: Science, Race, and Enlightenment in New Granada.* Translated by George Ciccariello-Maher and Don T. Deere. London: Rowman and Littlefield International, 2021.

Certeau, Michel de. *The Practice of Everyday Life.* Translated by Steven Rendall. Berkeley: University of California Press, 1984.

Cerutti Guldberg, Horacio. *Filosofía de la liberación latinoamericana.* Mexico City: Fondo de Cultura Económica, 1983.

Césaire, Aimé. *Discourse on Colonialism.* Translated by Joan Pinkham. New York: Monthly Review Press, 1972.

Chaunu, Pierre. *L'expansion européenne: Du XIIIᵉ au XVᵉ siècle.* Paris: Presses Universitaires de France, 1969.

Conley, Tom. "Early Modern Literature and Cartography: An Overview." In *Cartography in the European Renaissance*, vol. 3 of *The History of Cartography*, edited by David Woodward, 401–11. Chicago: University of Chicago Press, 2007.

Cotton, John. "God's Promise to His Plantation." Boston, 1630.

Coulthard, Glen. *Red Skin, White Masks: Rejecting the Colonial Politics of Recognition.* Minneapolis: University of Minnesota Press, 2014.

Craib, Raymond B. *Cartographic Mexico: A History of State Fixations and Fugitive Landscapes.* Durham, NC: Duke University Press, 2004.

Davenport, Frances Gardiner, ed. "The Bull *Inter Caetera* (Alexander VI.) May 4, 1493." In *European Treaties Bearing on the History of the United States and Its Dependencies to 1648*, 71–78. Washington, DC: Carnegie Institute of Washington, 1917.

Davenport, Frances Gardiner, ed. "Treaty Between Spain and Portugal Concluded at Tordesillas, June 7, 1494." In *European Treaties Bearing on the History of the United States and Its Dependencies to 1648*, 84–100. Washington, DC: Carnegie Institute of Washington, 1917.

Davis, Benjamin P. *Choose Your Bearing: Édouard Glissant, Human Rights, and Decolonial Ethics.* Edinburgh: Edinburgh University Press, 2023.

Deere, Don Thomas. "Coloniality and Disciplinary Power: On Spatial Techniques of Ordering." *Inter-American Journal of Philosophy* 10, no. 2 (2019): 25–42.

Deere, Don Thomas. "The Spacing of Decolonial Aesthetics." *Journal of World Philosophies* 5, no. 1 (2020): 89–98.

Deleuze, Gilles, and Félix Guattari. *A Thousand Plateaus: Capitalism and Schizophrenia.* Translated by Brian Massumi. Minneapolis: University of Minnesota Press, 1987.

Deleuze, Gilles, and Félix Guattari. *What Is Philosophy?* New York: Columbia University Press, 1996.

Derrida, Jacques. *Writing and Difference.* Translated by Alan Bass. Chicago: University of Chicago Press, 1978.

Descartes, René. *Meditations on First Philosophy: With Selections from the Objections and Replies.* Edited by John Cottingham. Cambridge: Cambridge University Press, 1986.

Drabinski, John. *Glissant and the Middle Passage.* Minneapolis: University of Minnesota Press, 2019.

Drabinski, John E. "Poetics of the Mangrove." In *Deleuze and Race*, edited by Arun Saldanha and Jason Michael Adams, 288–99. Edinburgh: Edinburgh University Press, 2013.

Dussel, Enrique. "Agenda para un diálogo filosófico Sur-Sur." In *Filosofías del Sur y descolonización*, 199–220. Buenos Aires: Editorial Docencia, 2014.

Dussel, Enrique. *Ethics of Liberation in the Age of Globalization and Exclusion.* Edited by Alejandro A. Vallega. Translated by Eduardo Mendieta, Nelson Maldonado-Torres, Yolanda Angulo, and Camilo Pérez-Bustillo. Durham, NC: Duke University Press, 2013.

Dussel, Enrique. *1492: El encubrimiento del otro: Hacia el origen del "Mito de la Modernidad."* La Paz: Plural Editores, 1994.

Dussel, Enrique. *The Invention of the Americas: Eclipse of "the Other" and the Myth of Modernity.* Translated by Michael D. Barber. New York: Continuum, 1995.

Dussel, Enrique. *Philosophy of Liberation.* Translated by Aquilina Martinez and Christine Morkovsky. Eugene, OR: Wipf and Stock, 2003.

Dussel, Enrique. *Posmodernidad y transmodernidad: Diálogos con la filosofía de Gianni Vattimo.* Mexico City: Universidad Iberoamericana, 1999.

Dussel, Enrique. *The Underside of Modernity: Apel, Ricouer, Rorty, Taylor, and the Philosophy of Liberation.* Translated and edited by Eduardo Mendieta. Atlantic Highlands, NJ: Humanities Press, 1996.

Dussel, Enrique. "World-System and 'Trans'-Modernity." Translated by Alessandro Fornazzari. *Nepantla: Views from the South* 3, no. 2 (2002): 221–44.

Elden, Stuart. *The Birth of Territory.* Chicago: University of Chicago Press, 2013.

Escobar, Jesús. "Toward an *urbanismo austríaco*: An Examination of Sources for Urban Planning in the Spanish Habsburg World." In *Early Modern Urbanism and the Grid: Town Planning in the Low Countries in International Context*, edited by Piet Lombaerde and Charles Van den Heuvel, 161–75. Turnhout, Belgium: Brepols, 2011.

Fanon, Frantz. *The Wretched of the Earth.* Translated by Richard Philcox. New York: Grove Press, 2004.

Foucault, Michel. *Abnormal: Lectures at the Collège de France, 1974–1975.* Translated by Graham Burchell. New York: Picador, 2004.

Foucault, Michel. *Discipline and Punish: The Birth of the Prison.* Translated by Alan Sheridan. New York: Vintage Books, 1995.

Foucault, Michel. *The History of Madness.* Edited by Jean Khalfa. Translated by Jonathan Murphy. London: Routledge, 2009.

Foucault, Michel. *The History of Sexuality.* Vol. 1, *An Introduction.* Translated by Robert Hurley. New York: Pantheon Books, 1978.

Foucault, Michel. "Of Other Spaces." Translated by Jay Miskowiec. *Diacritics* 16, no. 1 (1986): 22–27.

Foucault, Michel. *The Order of Things: An Archaeology of the Human Sciences.* New York: Random House, 1970.

Foucault, Michel. "Questions on Geography." In *Power/Knowledge: Selected Interviews and Other Writings 1972–1977*, edited by Colin Gordon, 63–77. New York: Pantheon Books, 1980.

Foucault, Michel. *Security, Territory, Population: Lectures at the Collège de France, 1977–1978.* Edited by Michel Senellart. Translated by Graham Burchell. New York: Picador, 2009.

Foucault, Michel. *Society Must Be Defended: Lectures at the Collège de France, 1975–1976.* Translated by David Macey. New York: Picador, 2003.

Foucault, Michel. "What Is Enlightenment?" In *The Foucault Reader*, edited by Paul Rabinow, translated by Catherine Porter, 32–50. New York: Pantheon Books, 1984.

Fraser, Valeria. *The Architecture of Conquest: Building in the Viceroyalty of Peru, 1535–1635*. New York: Cambridge University Press, 1990.

Galeano, Eduardo. *Open Veins of Latin America: Five Centuries of the Pillage of a Continent*. Translated by Cedric Belfrage. New York: Monthly Review Press, 1973.

Galli, Carlo. *Political Spaces and Global War*. Edited by Adam Sitze. Translated by Elisabeth Fay. Minneapolis: University of Minnesota Press, 2010.

Ganson, Barbara. *The Guaraní Under Spanish Rule in the Río de la Plata*. Stanford, CA: Stanford University Press, 2003.

Gasparini, Graziano. "The Laws of the Indies: The Spanish-American Grid Plan." *The New City* 1 (1991): 6–33.

Glissant, Édouard. *Caribbean Discourse*. Translated by J. Michael Dash. Charlottesville: University Press of Virginia, 1989.

Glissant, Édouard. *Faulkner, Mississippi*. Translated by Barbara Lewis and Thomas C. Spear. Chicago: University of Chicago Press, 1999.

Glissant, Édouard. *Introduction to a Poetics of Diversity*. Translated by Celia Britton. Liverpool, UK: Liverpool University Press, 2020.

Glissant, Édouard. *Philosophie de la Relation: Poésie en étendue*. Paris: Gallimard, 2009.

Glissant, Édouard. *Poetic Intention*. Translated by Nathanaël. Brooklyn: Nightboat Books, 2010.

Glissant, Édouard. *Poetics of Relation*. Translated by Betsy Wing. Ann Arbor: University of Michigan Press, 1997.

Glissant, Édouard. *Treatise on the Whole-World*. Translated by Celia Britton. Liverpool: Liverpool University Press, 2020.

Gómez, Fernando. *Good Places and Non-Places in Colonial Mexico: The Figure of Vasco de Quiroga*. Lanham, MD: University Press of America, 2001.

Gramsci, Antonio. "Some Aspects of the Southern Question." In *The Gramsci Reader*, edited by David Forgacs, 171–85. New York: NYU Press, 2000.

Grosfoguel, Ramón. "Developmentalism, Modernity, and Dependency Theory in Latin America." In *Coloniality at Large*, edited by Mabel Moraña, Enrique Dussel, and Carlos A. Jáuregui, 307–33. Durham, NC: Duke University Press, 2008.

Gualdrón Ramirez, Miguel. "To 'Stay Where You Are' as a Decolonial Gesture: Glissant's Philosophy of Antillean History in the Context of Césaire and Fanon." In *Memory, Migration, and (De)colonisation in the Caribbean and Beyond*, edited by Jack Daniel Webb, Roderick Westmaas, Maria del Pilar Kaladeen, and William Tantam, 133–51. London: University of London Press, 2020.

Hanke, Lewis. *All Mankind Is One: A Study of the Disputation Between Bartolomé de las Casas and Juan Ginés de Sepúlveda on the Religious and Intellectual Capacity of the American Indians*. DeKalb: Northern Illinois University Press, 1994.

Harley, J. B. *The New Nature of Maps: Essays in the History of Cartography*. Edited by Paul Laxton. Baltimore, MD: Johns Hopkins University Press, 2002.

Harvey, David. *The Condition of Postmodernity: An Enquiry into the Origins of Cultural Change*. Hoboken, NJ: Wiley-Blackwell, 1991.

Hegel, G. W. F. *Lectures on the Philosophy of History.* Translated by J. Sibree. London: G. Bell and Sons, 1914.

Heller-Roazen, Daniel. *The Enemy of All: Piracy and the Law of Nations.* New York: Zone Books, 2009.

Himmerich y Valencia, Robert. *The Encomenderos of New Spain, 1521–1555.* Austin: University of Texas Press, 1991.

Hobbes, Thomas. *Leviathan.* Edited by Richard Tuck. Cambridge: Cambridge University Press, 1996.

Hooker, Juliet. *Theorizing Race in the Americas: Douglass, Sarmiento, Du Bois, and Vasconcelos.* Oxford: Oxford University Press, 2019.

Hubbard, William. *The Happiness of a People in the Wisdom of Their Rulers Directing and in the Obedience of Their Brethren Attending unto What Israel Ought to Do.* Boston, 1676.

Huffer, Lynne. *Mad for Foucault: Rethinking the Foundations of Queer Theory.* New York: Columbia University Press, 2009.

Jáuregui, Carlos A., and David Solodkow. "Biopolitics and the Farming (of) Life in Bartolomé de las Casas." In *Bartolomé de las Casas, O.P.: History, Philosophy, and Theology in the Age of European Expansion,* edited by David Thomas Orique, OP, and Rady Roldán-Figueroa, 127–66. Leiden, Netherlands: Brill, 2019.

Kagan, Richard L. "A World Without Walls: City and Town in Colonial Spanish America." In *City Walls: The Urban Enceinte in Global Perspective,* edited by James D. Tracy, 117–54. Cambridge: Cambridge University Press, 2000.

Kant, Immanuel. *The Critique of Pure Reason.* Translated and edited by Paul Guyer and Allen Wood. Cambridge: Cambridge University Press, 1998.

Kant, Immanuel. "Perpetual Peace." In *Political Writings,* edited by H. S. Reiss, 93–130. New York: Cambridge University Press, 1991.

King, Tiffany Lethabo. *The Black Shoals: Offshore Formations of Black and Native Studies.* Durham, NC: Duke University Press, 2020.

Kotef, Hagar. "Movement." *Political Concepts: A Critical Lexicon,* no. 2 (2013). https://www.politicalconcepts.org/movement-hagar-kotef/.

Kotef, Hagar. *Movement and the Ordering of Freedom: On Liberal Governances of Mobility.* Durham, NC: Duke University Press, 2015.

Las Casas, Bartolomé de. "El Tratado de 'Las Doce Dudas.'" In *Obras Completas,* 11.2, edited by J. B. Lassegue, O.P., 11–224. Madrid: Alianza Editorial, 1992.

Lee, Richard E., ed. *The Longue Durée and World Systems Analysis.* Albany: SUNY Press, 2012.

Lefebvre, Henri. *The Production of Space.* Translated by Donald Nicholson-Smith. Oxford: Blackwell, 1991.

Levinas, Emmanuel. *Totality and Infinity: An Essay on Exteriority.* Translated by Alphonso Lingis. Norwell, MA: Kluwer, 1991.

Locke, John. *Second Treatise of Government.* Edited by C. B. Macpherson. Indianapolis, IN: Hackett, 1980.

Lugones, María. "Heterosexualism and the Colonial/Modern Gender System." *Hypatia* 22, no. 1 (2007): 186–209.

Lugones, María. *Pilgrimages/Peregrinajes: Theorizing Coalition Against Multiple Oppressions.* Lanham, MD: Rowman and Littlefield, 2003.

Lyotard, Jean-François. *The Postmodern Condition: A Report on Knowledge.* Translated by Geoffrey Bennington and Brian Massumi. Minneapolis: University of Minnesota Press, 1984.

Maffie, James. *Aztec Philosophy: Understanding a World in Motion.* Boulder: University of Colorado Press, 2013.

Mair, John. *In Secundum Sententiarum.* Paris: 1510.

Maldonado-Torres, Nelson. *Against War: Views from the Underside of Modernity.* Durham, NC: Duke University Press, 2008.

Maldonado-Torres, Nelson. "The Topology of Being and the Geopolitics of Knowledge: Modernity, Empire, Coloniality." *City* 8, no. 1 (2004): 29–56.

Marcos, Sylvia. *Taken from the Lips: Gender and Eros in Mesoamerican Religions.* Leiden, Netherlands: Brill Academic, 2006.

McKittrick, Katherine. *Demonic Grounds: Black Women and the Cartographies of Struggle.* Durham, NC: Duke University Press, 2006.

Mignolo, Walter D. *The Darker Side of the Renaissance: Literacy, Territoriality, and Colonization.* Ann Arbor: University of Michigan Press, 2003.

Mignolo, Walter D. "Preamble: The Historical Foundation of Modernity/Coloniality and the Emergence of Decolonial Thinking." In *A Companion to Latin American Literature and Culture,* edited by Sara Castro-Klaren, 12–32. Hoboken, NJ: Wiley Blackwell, 2008.

Mills, Charles W. *The Racial Contract.* Ithaca, NY: Cornell University Press, 1997.

Mundigo, Axel I., and Dora P. Crouch. "The City Planning Ordinances of the Laws of the Indies Revisited." *Town Planning Review* 48, no. 3 (1977): 247–68.

Mundy, Barbara. *The Mapping of New Spain: Indigenous Cartography and the Maps of the Relaciones Geográficas.* Chicago: University of Chicago Press, 1996.

Nebrija, Antonio de. *Gramática de la lengua castellana.* Madrid: Editorial Nacional, 1980.

Nemser, Daniel. *Infrastructures of Race: Concentration and Biopolitics in Colonial Mexico.* Austin: University of Texas Press, 2017.

Neuwirth, Steven D. "The Images of Place: Puritans, Indians, and the Religious Significance of the New England Frontier." *American Art Journal* 18, no. 2 (1986): 42–53.

Nichols, Robert. *Theft Is Property! Dispossession and Critical Theory.* Durham, NC: Duke University Press, 2020.

O'Gorman, Edmundo. *La invención de América: Investigación acerca de la estructura histórica del Nuevo Mundo y del sentido de su devenir.* Mexico City: Fondo de Cultura Económica, 2006.

Ortega, Mariana. *In-Between: Latina Feminist Phenomenology, Multiplicity, and the Self.* Albany: SUNY Press, 2016.

Padrón, Ricardo. "Mapping Plus Ultra: Cartography, Space, and Hispanic Modernity." *Representations* 79, no. 1 (2002): 28–60.

Padrón, Ricardo. *The Spacious Word: Cartography, Literature, and Empire in Early Modern Spain.* Chicago: University of Chicago Press, 2004.

Pagden, Anthony. *Spanish Imperialism and the Political Imagination: Studies in European and Spanish-American Social and Political Theory (1513–1830).* New Haven, CT: Yale University Press, 1990.

Pinet, Simone. "Literature and Cartography in Early Modern Spain: Etymologies and Conjectures." In *The History of Cartography*, vol. 3, *Cartography in the European Renaissance*, edited by David Woodward, 469–76. Chicago: University of Chicago Press, 2007.

Pitts, Andrea. *Nos/Otras: Gloria E. Anzaldúa, Multiplicitous Agency, and Resistance.* Albany: SUNY Press, 2021.

Portuondo, María M. *Secret Science: Spanish Cosmography and the New World.* Chicago: University of Chicago Press, 2009.

Prebisch, Raúl. *The Economic Development of Latin America and Its Principal Problems.* New York: United Nations, 1950.

Quijano, Aníbal. "Coloniality of Power, Eurocentrism, and Latin America." *Nepantla: Views from the South* 1, no. 3 (2000): 533–80.

Rama, Ángel. *The Lettered City.* Translated by John Charles Chasteen. Durham, NC: Duke University Press, 1996.

Rivera, Omar. *Andean Aesthetics and Anticolonial Resistance.* London: Bloomsbury, 2022.

Rojas-Mix, Miguel A. *La Plaza mayor: El Urbanismo, instrument de dominio colonial.* Barcelona: Muchnik Editores de Idiomas Vivientes, 1978.

Rorty, Richard. *Philosophy and the Mirror of Nature.* Princeton, NJ: Princeton University Press, 1979.

Rose-Redwood, Reuben S. "Genealogies of the Grid: Revisiting Stanislawski's Search for the Origin of the Grid-Pattern Town." *Geographical Review* 98, no. 1 (2008): 42–58.

Rousseau, Jean-Jacques. "Discourse on the Origins and Foundations of Inequality Among Men or Second Discourse." In *The Discourses and Other Early Political Writings*, edited by Victor Gourevich. Cambridge: Cambridge University Press, 1997.

Safier, Neil. *Measuring the New World: Enlightenment Science and South America.* Chicago: University of Chicago Press, 2008.

Santos, Boaventura de Sousa. "Beyond Abyssal Thinking: From Global Lines to Ecologies of Knowledge." In *Epistemologies of the South: Justice Against Epistemicide*, 118–35. New York: Routledge, 2016.

Sarmiento, Domingo F. *Conflicto y armonías de las razas en América.* Madrid: Akal Press, 2018.

Sarmiento, Domingo F. *Facundo o Civilización y barbarie.* Caracas: Biblioteca Ayacucho, 1986.

Sarmiento, Domingo F. *Facundo: Or, Civilization and Barbarism.* Translated by Mary Peabody Mann. New York: Penguin, 1998.

Schmitt, Carl. *The Concept of the Political.* Translated by George Schwab. Chicago: University of Chicago Press, 2007.

Schmitt, Carl. *The Nomos of the Earth in the International Law of Jus Publicum Europaeum.* Candor, NY: Telos Press, 2006.

Schutte, Ofelia. *Cultural Identity and Social Liberation in Latin American Thought.* Albany: SUNY Press, 1993.

Scott, James Brown. *The Catholic Conception of International Law: Francisco de Vitoria, Founder of Modern Law of Nations.* Clark, NJ: Lawbook Exchange, 2014.

Sealey, Kris. *Creolizing the Nation.* Chicago: Northwestern University Press, 2020.

Sealey, Kris. "When Heads Bang Together: Creolizing and Indigenous Identities in the Americas." *Puncta: Journal of Critical Phenomenology* 5, no. 4 (2023): 167–90.

Sepúlveda, Juan Ginés de. "Democrates Part Two, on the Just Reasons for the War Against the Indians." In *Sepúlveda on the Spanish Invasion of the Americas: Defending Empire, Debating Las Casas,* edited by Luke Glanville, David Lupher, and Maya Feile Tomes, 65–190. Oxford: Oxford University Press, 2023.

Sharpe, Christina. *In the Wake: On Blackness and Being.* Durham, NC: Duke University Press, 2016.

Silva, Grant. "'The Americas Seek Not Enlightenment but Liberation': On the Philosophical Significance of Liberation for Philosophy in the Americas." *The Pluralist* 13, no. 2 (2018): 1–21.

Simpson, Audra. *Mohawk Interruptus: Political Life Across the Borders of Settler States.* Durham, NC: Duke University Press, 2014.

Simpson, Leanne Betasamosake. *As We Have Always Done: Indigenous Freedom Through Radical Resistance.* Minneapolis: University of Minnesota Press, 2017.

Simpson, Lesley Byrd. *The Encomienda in New Spain: The Beginning of Spanish Mexico.* Berkeley: University of California Press, 1982.

Simpson, Lesley Byrd. *The Laws of Burgos of 1512–1513: Royal Ordinances for the Good Government and Treatment of the Indians.* San Francisco, CA: John Howell, 1960.

Sloterdijk, Peter. *Globes: Spheres II.* Translated by Wieland Hoban. Cambridge, MA: Semiotext(e), 2014.

Sloterdijk, Peter. *In the World Interior of Capital: Towards a Philosophical Theory of Globalization.* Translated by Wieland Hoban. Malden, MA: Polity Press, 2013.

Stanislawski, Dan. "The Origin and Spread of the Grid-Pattern Town." *Geographical Review* 36, no. 1 (1946): 105–20.

Stoler, Ann Laura. *Race and the Education of Desire: Foucault's History of Sexuality and the Colonial Order of Things.* Durham, NC: Duke University Press, 1995.

Subcomandante Marcos, Indigenous Revolutionary Clandestine Committee, EZLN. "Fourth Declaration of the Lacandon Jungle." January 1996. https://schoolsforchiapas.org/wp-content/uploads/2014/03/Fourth-Declaration-of-the-Lacandona-Jungle-.pdf.

Subcomandante Marcos. "Tomorrow Begins Today." In *Our Word Is Our Weapon,* edited by Juana Ponce de León, 99–105. New York: Seven Stories Press, 2001.

Subirats, Eduardo. *El Continente vacío.* Mexico City: Siglo Veintiuno Editores, 2012.

Tejeira-Davis, Eduardo. "Pedrarías Davila and His Cities in Panama, 1513–1522: New Facts on Early Spanish Settlement in America." *Jahrbuch für Geschichte von Staat Wirtschaft und Gesellschaft Lateinamerikas* 33 (1996): 27–61.

Tyler, S. Lyman. ed., *Spanish Laws Concerning Discoveries, Pacifications and Settlements Among the Indians.* Salt Lake City: American West Center, University of Utah, 1980.

Vallega, Alejandro. *Latin American Philosophy from Identity to Radical Exteriority.* Bloomington: Indiana University Press, 2014.

Vattimo, Gianni. *The End of Modernity: Nihilism and Hermeneutics in Postmodern Culture.* Translated by Jon R. Synder. Baltimore, MD: Johns Hopkins University Press, 1988.

Vera, Kim Benita. "From Papal Bull to Racial Rule: Indians of the Americas, Race, and the Foundations of International Law." *California Western International Law Journal* 42, no. 2 (2012):453–72. https://scholarlycommons.law.cwsl.edu/cwilj/vol42/iss2/12.

Veronelli, Gabriela Alejandra. "A Coalitional Approach to Theorizing Decolonial Communication." *Hypatia* 31, no. 2 (2016): 404–20.

Vieira, Antônio. *Sermão do espírito santo.* 1657.

Vitoria, Francisco de. *Vitoria: Political Writings.* Edited by Anthony Pagden and Jeremy Lawrance. Cambridge: Cambridge University Press, 1991.

Viveiros de Castro, Eduardo. *The Inconstancy of the Indian Soul: The Encounter of Catholics and Cannibals in 16th-Century Brazil.* Translated by Gregory Duff Morton. Chicago: Prickly Paradigm Press, 2011.

Von Vacano, Diego. "Paradox of Empire: Las Casas and the Birth of Race." In *The Color of Citizenship: Race, Modernity, and Latin American/Hispanic Political Thought,* 26–55. New York: Oxford University Press, 2012.

Wallerstein, Immanuel. *The Modern World-System I: Capitalist Agriculture and the Origins of the European World-Economy in the Sixteenth Century.* Berkeley: University of California Press, 2011.

Weheliye, Alexander G. *Habeas Viscus: Racializing Assemblages, Biopolitics, and Black Feminist Theories of the Human.* Durham, NC: Duke University Press, 2014.

Woodward, David. "Maps and the Rationalization of Geographic Space." In *Circa 1492: Art in the Age of Exploration,* edited by Jay A. Levenson, 83–87. Washington, DC: National Gallery of Art, 1991.

Wynter, Sylvia. "1492: A New World View." In *Race, Discourse, and the Origin of the Americas: A New World View,* edited by Vera Lawrence Hyatt and Rex Nettleford, 20–28. Washington, DC: Smithsonian Institution Press, 1995.

Wynter, Sylvia. "On How We Mistook the Map for the Territory, and Re-Imprisoned Ourselves in Our Unbearable Wrongness of Being, of *Désêtre.*" In *Not Only the Master's Tools,* edited by Lewis R. Gordon and Jane Anna Gordon, 107–69. London: Routledge, 2006.

Wynter, Sylvia. "The Pope Must Have Been Drunk, the King of Castile a Madman: Culture as Actuality, and the Caribbean Rethinking Modernity." In *Reordering of Culture: Latin America, the Caribbean and Canada in the Hood,* edited by Alvina Ruprecht and Cecilia Taiana, 17–41. Montreal: McGill-Queen's University Press, 1995.

Wynter, Sylvia. "Unsettling the Coloniality of Being/Power/Truth/Freedom: Towards the Human, After Man, Its Overrepresentation—an Argument." *CR: The New Centennial Review* 3, no. 3 (2003): 257–337.

Zambrana, Rocío. *Colonial Debts: The Case of Puerto Rico.* Durham, NC: Duke University Press, 2021.

Zorrilla, Victor. "Just War in Las Casas's Tratado de las doce dudas." In *Bartolomé de las Casas, O.P.: History, Philosophy, and Theology in the Age of European Expansion,* edited by David Thomas Orique, OP, and Rady Roldán-Figueroa, 243–59. Leiden, Netherlands: Brill, 2019.

Index

Heidegger, Martin, 65, 115n37
heterarchic: account, 104n26; conception of power, 6, 113n7; method, 7; logic of coloniality, 16, 104n26; reading, 6, 7
heterarchy, 85, 105n37, 118n8
heterotopias, 7, 18–19, 25, 27, 109n48; Americas as, 92; Borges and, 3; colonies as, 19, 21, 28; Foucault on, 103n11, 106n4, 107n13; grid as, 10; Indian town as, 24; of order, 86, 109n48
hierarchy, 58, 84, 93, 104n26; global, 58; of labor, 6; natural, 35; ontological, 64
Hobbes, Thomas, 110n1
Horkheimer, Max, 74
Hubbard, William, 47

identity, 68–69, 90–94; errantry and, 93; European, 92; logic of, 46
imperialism, 45, 67, 111n20
inconstancy, 23–24, 26
Indigeneity, 83, 85
Indigenous: communities, 19; cultures, 24; death, 85; dispossession, 84, 110n2; enclosures, 96; labor, 45, 87; land, 46, 53, 83; languages, 24, 88; nations, 38, 50; populations, 4, 32, 88; reason, 38; resistance, 12, 45–46, 48, 53, 83–85, 99, 118n5; rights, 37; self-determination, 111n18; societies, 87; sovereignty, 45; space, 10, 34, 37, 43, 45, 50, 52–53, 84; spatial distribution, 40, 42–43, 58–59; subjectivity, 5, 34; thought, 12; urban planning, 107n9; way of life, 24, 50; worldviews, 4
Indigenous peoples, 22, 24, 29, 31, 34, 36–38; language of Plantation and, 95; space and, 109n48; violence against, 37, 45. *See also* Native peoples
Indigenous space, 34, 45, 84
Indigenous subjects, 22–23, 26, 31, 34, 46, 58, 96, 108n39
interpellation, 74, 99
irrué, 12, 82–83, 85, 94–95, 101
ius gentium, 37–38, 44

Jesuits: in Brazil, 23–24, 49; in Paraguay, 19, 107n13
justice, 5, 62, 69, 73
just war, 37–38, 45, 112n41

Kant, Immanuel, 7, 9, 105n33, 110n5, 112n38
King, Tiffany Lethabo, 85
knowledge, 1–3, 5–7, 63–65, 69, 79, 81, 98, 113n7, 114n14; cartographic, 29; coloniality and, 6, 20; episteme of, 10; European, 32, 60, 64; geography of, 74, 78; geopolitics of, 5, 58, 63, 114n16; Glissant and, 87, 90, 94–96; grid and, 2; heterotopias and, 24–25; imperial, 102; invention and, 16; *irrué* and, 83; ordering of, 9, 17; subaltern, 104n21; subjugated, 11, 66, 78

labor, 20–21, 23, 41, 46, 50–52, 92, 104n20; compulsory, 108n34; Indigenous, 45, 87; law and, 40; relations, 104n25
land, 29–30, 34, 43–53, 66, 82–85, 92–93; appropriation, 36, 41, 43, 112n38; Indigenous/Native, 83–85; Indigenous relation to, 46, 53; liquefaction of, 38–41, 83; papal bull of donation and, 5, 36; property and, 33, 110n1; right to, 4, 35, 50
langage, 88–89, 96–100
language, 74, 77, 89, 97–99; of the center, 116n51; coloniality of, 107n10, 116n52; common, 77, 88, 96, 100; of the master, 74; mythological, 40; ordering of, 3; organization of, 17; pan-language, 103n7; Plantation and, 95, 99; rights-based, 46; universal, 3; written, 35
Las Casas, Bartolomé de, 35, 37, 62, 64, 111nn17–18, 112n41
Latin America, 9, 68–69; Caribbean coast of, 95; coloniality of space and, 12; dependency theory in, 67–68, 104n25; Dussel and, 115n37; independence period in, 17, 28; positivism in, 115n30
Laws of Burgos, 16, 22, 24–25, 29, 58–59
Lefebvre, Henri, 8, 105n34
Levinas, Emmanuel, 68–69, 115n37
Locke, John, 10, 34, 48–50, 66, 92, 110n1, 112n49. *See also* land; subjectivity: industrious
Lugones, María, 7–9, 11, 22, 34, 46, 51–52, 83, 102; creolization and, 118n3; "Heterosexualism," 104n20; *Pilgrimages/Peregrinajes*, 105n29; tactical strategies, 105n28. *See also* hangout; resistance
Lyotard, Jean-François, 73, 116n50, 117n55

positivism, 67, 115n30
postmodern thought, 67, 74–75, 117n56
power, 5–7, 9, 16–18, 21–22, 25, 51–52, 58, 63, 75–76, 113n7, 114n14; Americas as laboratory of, 86; apparatuses of, 24–25; of the center, 62; colonial, 43, 45, 85, 98, 107n17; coloniality of, 108n23; disciplinary, 16, 19–20, 22, 28–29, 61; Foucault on, 105n37, 108n23; global, 1, 6; grid and, 2, 24; heterarchy of, 118n8; map of, 34, 44–46, 50–51, 83; maritime, 29; material space of, 103n11; ordering of, 3, 9; of philosophy, 68; productivity of, 23; of reason, 105n31; relations, 6–7, 9, 21, 45, 97, 104n25, 113n7; sacralization of, 104n26; Spanish, 27; systems of, 81; techniques of, 16, 19–20, 32, 106n7, 107n17; temporal, 37
primitive accumulation, 39, 110n2
property, 110n1; Locke's theory of, 10, 34, 49, 92, 110n1; private, 38, 46, 50, 83; relations, 33, 110n2
proxemia, 59, 113n3
proximity, 21, 23–24, 26, 59, 113n3
psychiatry, 61, 77
Puritans, 47–50, 53, 112n47, 112n49; settler colonialism of, 10

queer theory, 117n60
Quijano, Aníbal, 6, 20, 22, 104n20, 104n25

race, 2, 25, 35–36, 40, 51, 104n25, 107–8n19; coloniality and, 6, 22; construction of, 17, 20; geography of, 30; Las Casas and, 111n17; mixing, 17, 30–31; moral geography of, 33, 35, 50; spatialization of, 17, 29, 35, 106n7, 110n8
racialization, 35, 110n15, 111n33; of geography, 50; of space, 10, 24, 26, 30, 36, 39, 64, 106n7
Rama, Ángel, 9, 19, 107n10, 116n46
rayas, 43, 110n13
reason, 6, 35, 80, 105n32; abyssal, 36; Caribbean, 89, 95, 98; of the center, 73, 77–79, 116n51; colonial, 83, 98; coloniality of space and, 51; colonizer and, 44; community of, 74; emancipatory, 71, 74; Eurocentric, 74, 76; European, 59, 79; Foucault on, 76–77; geography of, 72, 74,

80; instrumental, 59, 66, 72, 75, 99, 116n43; liberatory, 66–67, 80; modern, 11, 71, 73, 75–78, 83, 102; natural, 38, 45, 49–50; norms of, 37; of the other, 12, 71–73, 75, 77, 80, 98–99; of the periphery, 62–64, 73; pluriversal, 11–12, 67, 77–78, 98; powers of, 7, 105n31; spatiality of, 80; territoriality and, 37; transmodern, 78, 98; transmodernity and, 73, 86; Western, 75
reducciones, 87, 96
refusal, 10, 35, 46, 53, 84, 102; right to, 45–46, 53
relation, 34, 89–96, 98, 100–102; center-periphery, 74; colonial, 83; decolonization as, 93; Glissant on, 12; horizontal, 92–93; human, 88; identity, 93–94; Indigenous, 43, 46, 53; land-property, 110n1; networks of, 78, 90; of nonrelation, 58–60, 64, 79; Plantation and, 96; opacities in, 84; poetics of, 88, 90, 98, 101; property, 33; resistant, 87; rhizomatic, 119n24; space of, 2, 86–87; transmodernity and, 12, 59
representation, 2–3, 95; overrepresentation, 111n15; perfect, 1
resistance, 6–8, 11, 23, 47, 61, 83–87, 94–102, 109n48, 118nn7–9; Black, 85, 118n5; Caribbean, 91; to coloniality of space, 11, 120n49; heterotopias and, 18; Indigenous, 46, 48, 53, 83, 85, 102, 118n5; *irrué*, 85; language and, 89; to map of power, 51; mestizo, 84; poetics of, 94; practices of, 11, 19; right to, 45; space of, 42; spatial, 35, 46, 107n15; spatiality and, 105n29; thought and, 63
rhizome, 90–91, 93, 119n22
right to travel, 34, 37–38, 49
rootedness, 47, 83, 90–91
Royal Ordinances on City Planning, 16, 25, 27, 108n36, 108n39

Santos, Boaventura de Sousa, 6, 36, 104n17
Sarmiento, Domingo F., 30–31, 39, 42, 110n5, 116n46. *See also* pampa
Sartre, Jean-Paul, 65, 114n19
Schmitt, Carl, 39–44, 110n13, 111n20, 111n33. *See also* nomos
Schutte, Ofelia, 68–69, 115n25, 115n27
Sealey, Kris, 84, 118n3, 118n5, 118n9

secret science, 1, 103n2
Sepúlveda, Juan Ginés de, 35, 37, 58–59, 62, 64, 113
settlement, 5, 10, 34, 38–39, 58; colonizer and, 52, 92; extraction and, 110n2; norms of, 50; property as enclosed, 92; Puritans and, 47–48; rhizomatic subjectivity and, 101; Spanish colonial, 25; traveler and, 91. *See also* movement
settler, the, 10, 33, 45, 47–48, 50–51, 85, 91; Spanish, 58; white, 47
Simpson, Audra, 53, 113n67
Simpson, Leanne Betasamosake, 83
slavery, 20–21, 61, 87, 107–8n19; natural, 35–36; racialized, 108n34
slave trade, 61, 88, 111n15
Sloterdijk, Peter, 39
smooth space, 30, 42–44, 85, 109n52, 111n27, 119n26, 119n30; rhizome and, 93; striation of, 39–40, 42–44, 47
solipsism, 65, 114n19
South-South dialogue, 70, 72, 78, 117n63
sovereignty, 23, 27–28, 38–39, 45, 50; nested, 53; Peruvian, 111n18
Spain, 27, 29, 36, 45, 96, 111n18, 112n41; Caribbean islands and, 118n16; gridded space and, 18; Portugal and, 5, 39
Spanish crown, 23, 25
Spanish empire, 1, 29
spatiality, 16, 50–52, 103n11, 109n48; of abyssal lines, 37; alternative, 102; Andean, 112n57; colonial, 112n57; of colonial conquest, 46; of coloniality, 58; of the grid, 26; Lugones on, 105n29; of oppression, 51, 53; of proximity, 59; racially coded, 50; of reason, 80; sovereign, 27; of the traveler, 45
spatial order, 9, 31, 39–41, 43
spatial ordering, 9, 16–17, 32, 41, 44, 57, 104n25
spatial turn, 105n34
state formation, 91, 109n44, 119n26
striated space, 42–44, 47, 61, 85, 90, 109n52, 111n27, 119n30; apparatus of capture and, 39
striation, 43, 51, 93; of American space, 45; modernity and, 101; of smooth space, 39–40, 44; of space, 29, 42
subjection, 16, 102

subjectivity, 10, 19, 21, 27, 29, 84, 87–88, 92, 115n38; African, 5; Black, 85; Caribbean, 107n15; cellular, 96; coloniality of space and, 51; of Cortés, 11; criollo, 5; decentered, 79; inculcation of, 22, 59; Indigenous, 5; industrious, 49–50, 105n32; mestizo, 5; ordering and, 2, 7, 19, 51; resistant, 52; rhizomatic, 101; self-policing, 20; space and, 2, 16; spatialization of, 34; temporality of, 8. See also *ego cogito*; *ego conquiro*
subsumption, 89; Marxian, 74; of the other, 93
suffering, 88–89, 99

temporalization, 8, 79, 115n28
terra nullius, 36, 83–84
territoriality, 37, 40
territory, 2, 29, 32, 36, 62, 91–93; coloniality of the state and, 83; constructions of, 33; history of, 105n34; liquefaction of, 40; relationality to, 46; rhizomatic subjectivity and, 101; right to, 35; sovereign, 27–28
totality, 5, 58, 66, 69, 93; of being, 63; nontotalizing, 89
totalization, 8, 57–59, 76, 79; Borges and, 3; Dussel's critique of, 66, 93
transmodernity, 11–12, 59–60, 70–75, 77, 86, 97–98, 100, 117n63; coloniality and, 80
transparency, 88, 98
traveler, 33, 38, 45–47, 50–52, 85, 91
Treaty of Tordesillas, 5, 18

unitopia, 7–8
universality, 66, 70, 89–90, 97; of Caribbean modernity, 88; colonizer and, 52; Europe and, 8, 11, 59–60, 63, 92
uprooting, 11, 47, 82, 86, 88, 93, 95, 101
urban planning, 28, 107n9

Valladolid debates, 35
Vattimo, Gianni, 74–75, 117nn55–56
Viera, Antônio, 23
violence, 88, 91; care and, 21, 23–24; colonial, 6, 37, 67, 75, 77–78; conceptual, 48; conquest and, 66; of conquistadors, 108n39; evangelization and, 58; of emptying, 2, 18; of fungibility, 102; histories of, 116n52; of modernity, 11, 71, 76; Plantation and, 96; reason and, 74–75; sacrificial, 70–71